KT-418-853

39
Ways
to
Save
the
Planet

Tom Heap

WITNESS
BOOKS

1

Witness Books, an imprint of Ebury Publishing
20 Vauxhall Bridge Road, London SW1V 2SA

BBC Books is part of the Penguin Random House group of companies
whose addresses can be found at global.penguinrandomhouse.com

First published by Witness Books in 2021

www.penguin.co.uk

A CIP catalogue record for this book is available from the British Library

ISBN 9781785946974

Editorial Director: Albert DePetrillo
Assistant Editor: Daniel Sørensen
Project Editor: Bethany Wright
Design and typesetting: seagulls.net

Printed and bound in Great Britain by Clays Ltd, Elcograf S.p.A.

The authorised representative in the EEA is Penguin Random House
Ireland, Morrison Chambers, 32 Nassau Street, Dublin DO2 YH68

Penguin Random House is committed to a sustainable future for
our business, our readers and our planet. This book is made
from Forest Stewardship Council® certified paper.

Contents

FARMING

SOCIETY

TRANSPORT

BUILDINGS AND INDUSTRY

WASTE

As Governor of California, Arnold Schwarzenegger fought to pass the 2006 Global Warming Solutions Act and the low carbon fuel standard, and delivered greater protection of nature. Today, improving the environment is a core purpose behind both the Schwarzenegger Institute for State and Global Policy at the University of Southern California and the Schwarzenegger Climate Initiative. Every year he hosts the Austrian World Summit, bringing together politicians, business people and thought leaders to tackle climate change.

Foreword

'There is no fate but what we make for ourselves' isn't just a fantastic line from *Terminator* – it's a philosophy we can all use in our lives.

The *Terminator* franchise portrays the apocalypse of machines trying to control our lives. It depicts a frightening, dystopian, uninhabitable world. But here's the thing: the *Terminator* movies did not dwell on the hopelessness of the situation. No, they focused on human will and human hope. In fact, when Sarah Connor warned the world about the coming robot apocalypse, they put her in an insane asylum.

I think it is important that we channel optimistic thinking into the current environmental movement instead of just freezing people with doom and gloom that they don't really understand. Yes, there is reason for alarm. Yes, it is true that we see the problems getting bigger and bigger, from wildfires and floods to shellfish boiling in their shells to water that isn't safe to drink and air that isn't safe to breathe. Yes, we are in a crucial moment.

But I believe we need to also share the hopeful stories of heroes terminating pollution all over the world, creating jobs and helping future generations breathe easier.

That's why I love this book. It's right there in the title: *39 Ways to Save the Planet*. 39 solutions, 0 doom. You will learn about different visionaries all over the world who have been working without much fanfare to solve our pollution crisis.

This is exactly what we need to show the people that we are in control.

Because let's be honest: we need the people to create our clean energy future, we need them to solve the pollution crisis that kills more than 7 million people each year and we need them to be a part of our environmental crusade.

We keep waiting for the governments of the world to solve this problem, but what we really need is the people of the world to solve this problem. Every great movement has enlisted the people – the civil rights movement, the anti-apartheid movement, the women's suffrage movement and the Indian independence movement. The environmental movement is no different and that is why I think this is such an important book for this moment in time. With so much writing on the climate crisis focused on the crisis, this book is focused on the solutions the people can get behind.

Across the globe, enterprising individuals are finding solutions to the challenges in front of us. They are turning to nature, technology, classrooms and their own homes to create clean energy, reduce emissions and pull carbon from the air. You can't help but feel inspired by the innovators chronicled in this book. You will love reading about the computer scientist who is using artificial intelligence to build robots that can service wind turbines miles out at sea. Or the former consumer products executive who has developed an inexpensive and much safer nuclear power plant. Or the professor who is working to de-gas cows. Or the teacher in Zimbabwe who is fighting climate change by educating girls about sustainable farming. Each of the 39 case studies in this book is inspiring and provides the reader with optimism for our future.

It's time for all of us to stop talking to people about what they have to lose and start talking to people about what they have to win. We need to recycle our white flags, terminate our doom and gloom, and start building a real movement with a foundation of optimism and hope.

Don't know where to begin? The book you're holding right now is a good place to start. I know that you'll be both awed and inspired by the people and projects you'll read about here.

Because good things are happening. We can be the change, and we can save the world.

There is no fate but what we make for ourselves. So let's make ours a clean, green, healthy fate.

Introduction

Climate change is a real and present danger to human civilisation on this planet, yet this book could have been titled '39 Reasons to Be Cheerful' because it reveals the women and men delivering solutions.

During 25 years of reporting on the environment, countryside and science I have become unshakeably convinced of the severity of what we are facing but I often feel the stories of those who are solving this most wicked of problems are being ignored. This struck me as odd, not simply because there are inspiring and novel things to discover out there, but because solely focusing on the gloom is disempowering. In a war – and some say we need a war footing to defeat climate change – public morale is maintained by headlining success and sidelining setbacks. In the fight against climate change, we appear to be doing the reverse, leaving many people feeling anxious and helpless. Now, I am not recommending ignoring the bad news but let's at least hear and cheer the many things that are going right: that way they grow. That is the motive behind filling this book and the BBC Radio 4 series that it accompanies with untold stories of ingenuity and redemption.

In mid-2021, as I complete this book, I have never felt such a buzz surrounding the need to tackle climate change. This is obviously exciting but also surprising as the clamour for action has grown alongside the more immediate threat of the COVID-19 pandemic; you might have expected dealing with daily death tolls to have shunted climate threats back down the track. Why this hasn't happened would probably fill another book but here is a trio of positive thoughts. We have learnt greater respect for science, and scientists want climate action. Governments have enacted sweeping policies to limit the impact of the coronavirus so they now feel emboldened to protect people long term through the rollout of tougher rules and targets to cut carbon. The third is a more slow-burn phenomenon: business seems to have reached a tipping point of realisation that there is now more money to be made in solving climate change than ignoring it.

There is no exact science in how the ideas and innovators were chosen for *39 Ways to Save the Planet* but I guess there are a few themes. Firstly, I was determined not to ignore the big solutions that are already working, such as wind power, solar energy, electric cars and regenerative farming. The speedy growth of these will make some of the biggest dents in climate change and I believe we have found people who are pressing the accelerator. We feature some 'hero' scientists dwelling anywhere from suburban garages to experimental seaweed farms to hi-tech clean rooms, but the emphasis is on the readiness of their solutions. Stunning scientific breakthroughs, like functional nuclear fusion, would be great but there is no need or point in waiting for the invention of the magic bullet. Given the urgency of the climate problem, now is better than new. But this book is no mere tech fest. There are solutions to be found in the classroom, the courts and in our kitchens. The natural world also has so much to offer. So-called 'nature-based solutions' such as planting bamboo, seeding seagrass, rewetting peat or spreading biochar can all help to halt existing emissions and create carbon sinks.

Also included are solutions that some environmental voices reject, such as nuclear power, carbon capture, reduced impact logging or climate-friendly cows. Opposition stems from innate fear, suspicion that they are linked to existing 'bad actors' like the fossil fuel companies, belief they won't work or that they will excuse us from making tough choices in own lifestyles. So, this is why they are included. Fundamentally, cutting carbon emissions is such a massive challenge for a civilisation reared on ever greater fuel consumption that we need every tool in the box to do it. Or as Christoph Gebald, from CO_2-capturing company Climeworks, puts it: 'When discussing climate-friendly technologies, you should not be allowed to use the word "or". You should only be allowed to use the word "and".'

Insisting on only engaging with technologies that come top of the environmental virtue scale or companies with no oily blemishes in their history risks striving for the perfect while blocking delivery of the good. We don't have time for that. And when it comes to expecting behavioural change to save the world … the track record is so dismal that I won't carry on playing a tune that hardly anyone is dancing to.

Most chapters contain or conclude with some estimate of the scale of an idea, for example how big a bite it could take out of the 50 billion tonnes of CO_2e (the 'e' means 'equivalent' and includes the impact of other greenhouse gases like methane, nitrous oxide and escaped refrigerants) we emit each year. This is very far from an exact science as it's so dependent on politics, investment and public engagement, which is why I've opted for the heading 'Desirable destination'. It is a goal that is within the physical boundaries of possibility and is plausible given sufficient will.

'Will' is the thing. Having met all these great people and seen what they are doing, I am completely convinced that we have the *ways* to bring carbon emissions down to net zero but do we have the *will*? Will isn't really abstract; it's proven by the laws and spending delivered by governments, the choices of business and the behaviour of all of us. Just because this isn't a book of low-carbon lifestyle tips shouldn't leave you feeling disempowered. You can help reach that 'desirable destination' through your vote, your work, your spending and your inspiration. It's within our grasp – let's do it.

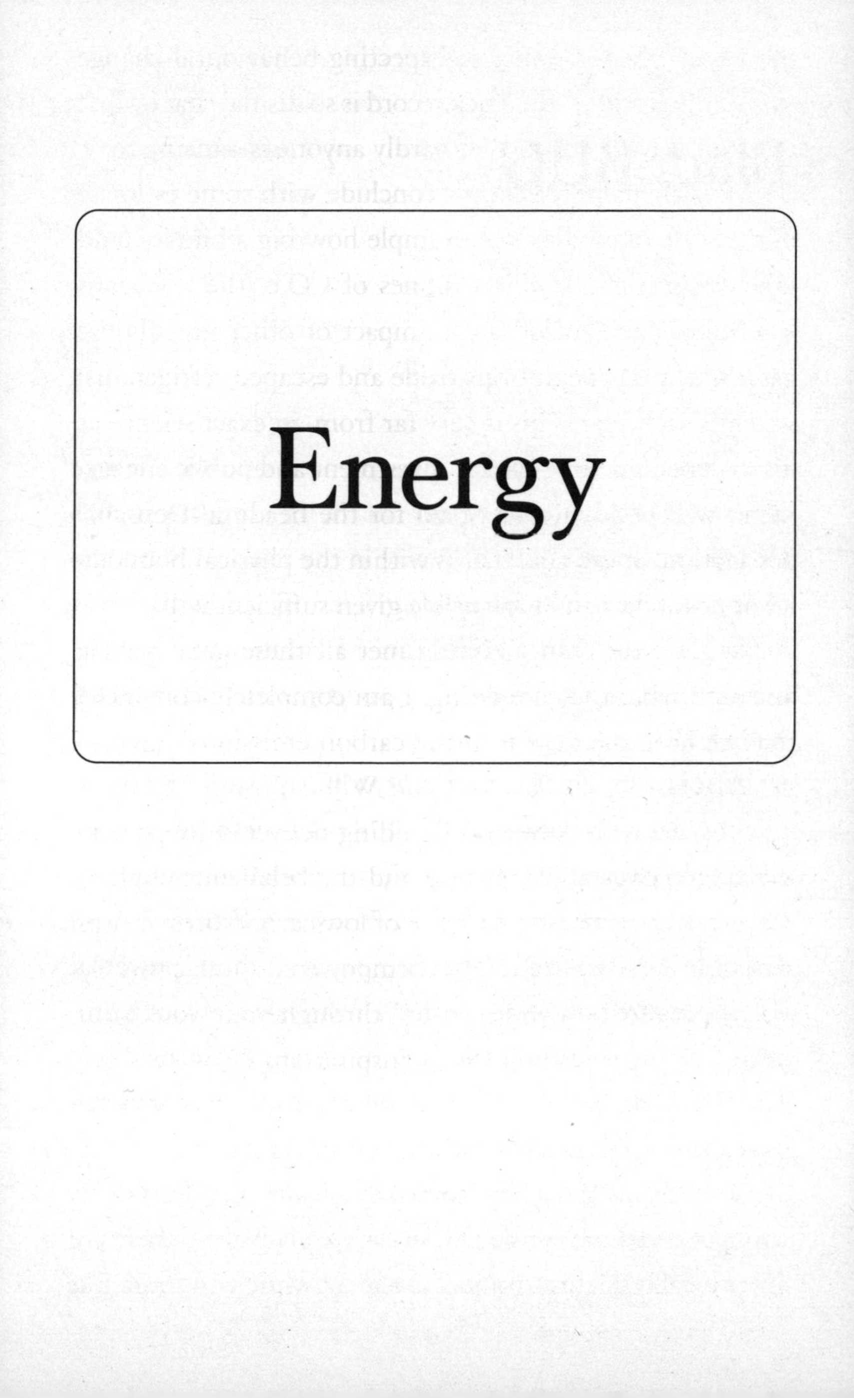

Energy

1

BladeBUG

In a workshop close to the Thames, a robotic ant the size of a small child is coming for me. With six suction cups for feet, nowhere is out of reach: it can even walk on the ceiling. Luckily, it is being controlled by an engineer, is moving slower than a soggy-limbed zombie and has the benign mission of massively expanding wind energy. This is BladeBUG.

Across the world, wind energy has grown rapidly in recent years with between 15 and 20 per cent more capacity added every year. Solar energy is on a similar curve and, for the planet, competing to be the leading source is a most desirable head-to-head. The biggest and most powerful windfarms are now at sea, with the relatively crowded European continent leading the way in development offshore. The UK alone has one third of the world's marine generating capacity, followed by Germany and China.

Windfarms at sea have some key advantages: the wind is stronger and more consistent, space is available and there are fewer local residents to object to a giant white windmill. But offshore windfarms have one sizeable problem: it's a hostile world out there for both man and machine so keeping everything up and running is both perilous and expensive. Over the 25–50-year life of an offshore turbine, including both construction and operation, maintenance typically accounts for around 40 per cent of the bill. If you can bring down those costs, new sites become economic and you've got a windrush.

The most promising way to do this in an inhospitable environment is to remove the human element. Over the last few decades of space exploration, it's been realised that you can 'boldly go' a lot further, faster and cheaper without having to keep 80 kg of human alive for the ride. And that is why one of the brains behind BladeBUG once planned autonomous space missions for NASA. Sarah Bernardini, a professor of artificial intelligence, says, 'Working on a thin turbine blade 100 metres above the waves, many miles out to sea, is not for humans. The environment is life threatening, the communications are poor and the cost astronomical: like space.'

Wind turbines are pretty robust but all machines need some upkeep, especially those generating electricity from the raw elements. The tip of a turbine blade can travel at 320 km/h. Even rain at that speed becomes like a shower of shrapnel. The support tower is subject to enormous pressures from regular storms. Inspection and repair are vital and currently carried out by teams deployed from floating cranes below or rope access from above. Such crews and their kit are necessarily expensive and all the time the turbine is switched off it's not making money, often losing a further £10,000 per day.

The vision of BladeBUG's founder Chris Cieslak is to keep all the humans onshore. An autonomous vessel would leave port carrying the BladeBUG. On its approach to the turbines a drone would take off and make an initial inspection from the air, then it would return to the boat, collect a BladeBUG robot and ferry it to the work area. It's like a Russian-doll series of robots with one carrying another, carrying another. Once onsite, the BladeBUG can inspect the surface visually and with ultrasonic non-destructive testing; even the feet have electronic sensors to detect cracks. It can also carry different payloads such as drills, spanners, grinders and resin injectors to do any repair work required. Its six feet and articulated body allow it to reach pretty much anywhere, even inside the tower.

When costs come down as production matures, Chris Cieslak sees a future where every turbine would have its own BladeBUG – like a tick-bird on a rhinoceros: 'It would be a turbine's personal groom: an individual inspection and repair robot, which would detect and sort problems quickly. Like a paint chip on your car, it's far cheaper to intervene early than let the problem worsen.'

Alongside the London workshop, the BladeBUGs are being tested at the Offshore Renewable Energy Catapult's centre in Blyth, northeast England. It is supported by the Catapult Network, a government-backed scheme for promising new technologies. Sara Bernardini is convinced these robots will become mainstream by the mid-2020s but there are still real challenges for artificial intelligence. At the moment decision-making is shared between a remote human operator and the machine itself but Sarah Bernardini envisages a steady handover of power: 'BladeBUG needs to be able to adapt according to changing weather conditions, react to what problems they find and use reasoning, along with other robots, to work out the best plan. They also need to learn to improve performance. This is the cutting edge of artificial intelligence.'

The UK aims to quadruple its offshore wind capacity to 40 gigawatts by 2030, and many other countries, notably the USA and China, have huge expansion plans. Norway and Spain are developing floating turbines – that may sound strange but remember we've had floating drilling platforms for decades – and these could operate in much deeper water, further from land. The International Energy Agency (IEA) says offshore wind has the potential to generate 420,000 terawatt hours of electricity per year. This is more than 18 times global electricity demand today. That is an implausible goal but shows that offshore wind is one of our really big solutions to climate change. All these ambitions will be more viable with robotic maintenance.

Desirable destination

By 2050 the International Energy Agency says offshore wind could grow sufficiently to cut 5 per cent of our current greenhouse gas emissions. This would require a 60-fold increase in the electricity generated by offshore wind turbines.

How to get there

Bigger turbines: the latest generation of offshore wind turbines are huge and getting bigger year on year – 250 metres tall and with blades catching the wind from a 'swept area' five times the size of a football pitch.

Technology: BladeBUG and other robotic and remote sensing techniques are being developed to reduce the cost of installation and maintenance. Floating offshore wind platforms using methods learnt from floating deep-sea oil and gas rigs are being tested.

Rewiring: you need cables to get the power from the sea to our homes; in the UK, a coastal ring main has been suggested to prevent the seaside becoming a forest of pylons.

Distant and deep waters: the wind tends to be stronger and more consistent further out to sea, the spinning blades present less risk to coastal wildlife and you can deploy them over the horizon to be less visually intrusive.

Fringe benefits

More good jobs: building and deploying such a huge number of offshore wind turbines is a huge business opportunity and in many places could replace jobs being lost in the fossil fuel industries.

Fewer perilous jobs: robots can do more of the really hazardous roles where lives are at risk.

More sealife: the base of wind turbines can provide structure and habitat on the seabed for marine life, and many offshore windfarms become no-take zones, providing a fish and shellfish refuge.

2

Floating Solar

Solar power is booming. As the price comes down, deployment goes up and more rural areas are silver-plated as huge solar farms take root. This is good for increasing low-carbon energy but not everyone is happy.

'Real farms not solar farms!'

'Say no to mega sun farm!'

'Fields 4 Food!'

Banners and posters sprout beside country roads as the 'food v. fuel' debate morphs into an unease about these shiny splashes on the landscape and how they might even threaten food supply.

'I wanted to be in a business with a big mission. I want to develop renewable energy with no conflict with land use,' says Alexis Gaveau, co-founder of French floating-solar pioneers Ciel et Terre. 'The idea came when we were prospecting for potential solar sites near Marseilles. The land was very built up but then we noticed these flooded former quarries.'

Ciel et Terre is now one of the world's biggest players in floating photovoltaics – FPV in the trade – with offices in 11 countries, over 500 megawatts of installed capacity and more than 1,000 megawatts in the development pipeline. The idea is simple: solar panels are fixed to buoyant plastic rafts anchored to the lake bed or tethered to the shore. When the sun shines the electricity is cabled back to dry land. Making FPV systems tough enough to last at least 25 years yet cheap enough to compete with land-based solar farms is the trick.

Where land area is free, or very affordable, FPV will always struggle to compete on price. Installation using boats, floats, divers and specialist training is likely to be more expensive than the 'man with a van' equivalent on dry land. Also, some engineering and cabling specifications are higher to cope with a watery home. But FPV does have some advantages. The reservoir tends to cool the panels, making them a little more efficient as electrical resistance reduces with cold, and the reflectivity of water can improve performance too. Some of the early FPV sites in Europe, such as at the Queen Elizabeth II reservoir in west London, are on reservoirs owned by water companies, and here the surface shading caused by FPV can be an asset in helping to reduce the growth of problematic algae, which thrives in sunshine. They also help to reduce evaporation losses. Many cities, with a huge appetite for electricity and prohibitively expensive land, are giant waterside settlements built around river estuaries or historic docks: potential floating solar real estate.

The most fertile waters for floating solar are in Asia, especially China, Japan, South Korea and Singapore where land is at a premium but water abounds. The global growth rate for FPV is expected to be at least 20 per cent per year with two thirds of that coming from South and East Asia. South Korea, a very mountainous country, is deploying FPV on the tidal flats of Saemangeum. On its completion in 2025 it will be the world's biggest floating solar installation with a 2.1 gigawatt capacity – roughly twice the power of a typical nuclear power station. Ciel et Terre has floating solar farms on irrigation ponds in Taiwan and Japan, and on water treatment works in China, India and New Zealand. In the twenty-first century, more of our planet has to multitask.

Doing two things at once is the persuasive idea behind the latest place for solar to take dip: hydro-electric reservoirs. Once again, the idea is a natural fit. Hydropower plants already have good grid connections so less money is required for cabling. Hydroelectricity depends, in essence, on rainfall and river flow; often this is lower in the summer months so plants can't run at full capacity, but this is usually when the sun shines strongest so solar can help make up the shortfall. Reduced evaporation from covering the surface also leaves more water behind the dam to power the turbines. Strength in one generation system makes up for the weakness in the other. The potential of this hydro/solar combo is yielding some pretty impressive figures: the US Department of Energy recently estimated that solar arrays on the world's hydroelectric reservoirs could meet half of global electricity demand. That scale of deployment

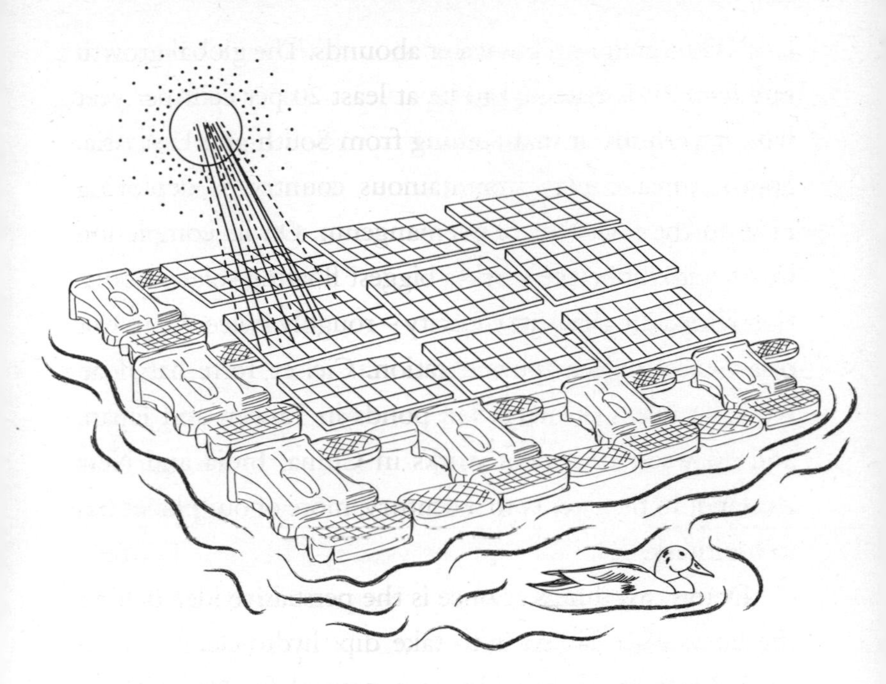

is highly unlikely but illustrates that FPV panels are not necessarily small fry.

Another inspiring, and possibly more attainable, projection comes from the European Joint Research Centre looking at Africa. Hydroelectric power delivers approaching 20 per cent of the continent's electricity, and in countries such as Ethiopia and Mozambique it is 90 per cent. Fifty new dams are under construction. The authors of the centre's report estimate that covering just 1 per cent of the continent's reservoirs with solar panels could double the existing capacity from those assets: 28 to 58 gigawatts. And here again the evaporation blocking and seasonal compensation offered by floating solar are other benefits. The authors also warn that climate change is interfering with the rainfall patterns in Africa, so spreading energy generation away from river flow is a sensible insurance policy. Slightly higher costs and engineering complexity remain but it feels like a good fit for a continent where many people live in energy poverty and the demand for electricity is rising sharply.

Obviously, the biggest frontier in floating solar is the water covering nearly three quarters of the globe: the sea. Alexis Gaveau says Ciel et Terre has existing marine projects and more on the way but stresses that it can be a tough environment: 'There are waves which can be big, winds which can be strong, the water is always salty and climate change is expected to make the weather more extreme. All of these are a challenge for an electrical installation.'

Nevertheless, giant flexible rafts are being deployed in sheltered coastal waters around Asia. Ciel et Terre has a raft in the sea around Taiwan that is bigger than 100 football pitches and is built to withstand 5-metre waves. The potential prize of open ocean solar is attracting other players who have their own experience in the offshore game: oil and gas companies with decades of hostile-environment engineering under their toolbelts and fish farmers who float for a living. In the next decade, the North Sea between the UK and continental Europe might see the first combined solar, wind and seaweed farm: a triple win to float my boat.

Desirable destination

Around 40 per cent of our greenhouse gas emissions comes from electricity generation and, theoretically, half of that – 20 per cent – could come from water-top solar. More likely, given frequently cheaper ground-based panels, would be a figure of 4 per cent.

How to get there

Policy: include water-top solar deployment as a condition for the development of new reservoirs.

Durability: prove the reliability and economic return from floating solar so it is attractive to energy companies.

Innovation: encourage experts in offshore engineering to cooperate on designing credible ocean solar.

Fringe benefits

- Less algae growth in drinking-water reservoirs.
- Less evaporation from hydroelectric lakes.
- With less land-based solar, more land left available for food or wildlife.

3

Gravitricity

There are times when an idea is so stunningly simple you wonder why it has taken so long to drop into your lap. Gravitational energy storage is right up there. We all know it takes energy to lift something – to move it against the force of gravity – that's why it is hard to bike uphill or lift weights in the gym. It's also common knowledge that falling releases that energy, hence speeding downhill or letting your dumbbells crash to the floor. So why not store electrical energy by lifting something heavy?

It is a truth universally acknowledged that an electricity grid in possession of a good proportion of renewable energy must be in want of energy storage. As a growing proportion of our electricity comes from the fickle wind and sun, the more ways we need to store it for use in the dark, cold and still times. There is a rapidly evolving ecosystem of new technologies to achieve just that: batteries of varying chemistry, compressed air, flywheels, hydrogen and gravity.

Dockside in Leith, on the northern shores of Edinburgh, sits what looks like the stub of a construction crane but is actually a heavy-duty frame to lift a total of 50 tonnes. It is the working prototype of Gravitricity, a Scottish company whose eventual plan is to store energy by raising and lowering blocks of iron ore weighing thousands of tonnes in mine shafts potentially thousands of metres deep. It is eyeing up commercial sites in Central Europe, northern Spain and South Africa: areas with a history of mining and a future in renewables.

Gravitricity's managing director, Charlie Blair, says there's a bit of 'back to the future' in the technology too. 'A grandfather clock is powered by gravitational energy. Its great height is there to provide a long drop for a weight. You lift it once a day and its gentle descent powers the mechanism. We'll be doing the same, only bigger.'

In fact, gravitational storage of energy is already very common; only it works with water and we call it hydroelectricity. Falling water held by a dam and released on demand spins turbines to generate electricity. In 1965, Cruachan, in the Scottish Highlands, became the first big reversible pumped storage system built in the world. It has a loch top and bottom; when there is spare electricity, water is pumped up, and when you need the power, you let it fall.

But there is a problem: it may not feel like it when you drop a hammer on your toe, but gravity is surprisingly weak. A lot of falling weight delivers surprisingly little energy. To power a typical 10-watt LED light bulb for an hour, you would need to lower 100 kilos – the weight of a large man – about 40 metres. A 10-metre drop would give you roughly the energy contained in a AA battery. That is why you

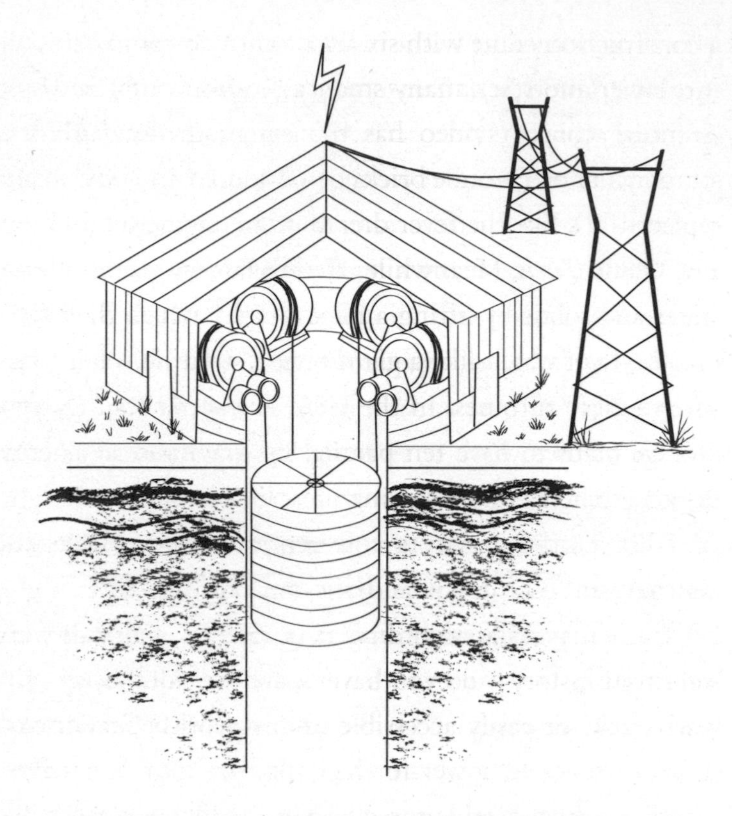

need massive weights and big drops combined with very heavy and very smart engineering. Gravitricity's winches must lift thousands of tonnes, work at varying speeds, cope with the consequent different stresses and generate electricity too. Bear in mind that the average lift weighs only 2–3 tonnes when fully laden. The winches are being built by the company behind the world's biggest cranes.

Gravitricity is not the only company inspired by Newton's apple. The town of Ticino sits amid a Swiss Alpine valley, but dwarfing the homes and offices is what looks like a construction crane with six arms. This 'power tower' will lift, lower and stack many smaller blocks around its base. A future concept video has the crane surrounded by a sturdy pillar of 35-tonne bricks being regularly removed and replaced. It's like the fever dream of an engineer raised on LEGO and *Tetris*. Meanwhile, Heindl Energy is combining water and solids by lifting a huge weight, which then falls on a body of water, like a giant piston, creating water pressure to drive turbines. In the USA, Advanced Rail Energy Storage plans to have ten parallel railway lines on a steep slope with hefty wagons going up and down.

With Earth's gravity a constant, the key to success is getting the costs down and the flexibility up. The right approach may vary according to a region's geography or industrial history – do you have 'spare' hillsides to lay railways tracks or easily accessible ex-mineworks? Gravitricity promises to store power for half the price of current lithium-ion batteries and it reckons its system will be able to go from zero to full power in less than a second. This speed is needed for what is known as 'grid balancing': the ability to match fluctuating demand, as people switch on kettles or plug in their electric car, with supply. National grids have become extremely adept at this fine art over the years but have often relied on being able to turn the burners up and down on gas-fired power stations. In a low-carbon world, that option is both less desirable and less available, so having climate-friendly ways of having electricity available at the flick of a switch is both critical and valuable.

Charlie Blair of Gravitricity also sees huge scope for gravity storage in poorer countries with emerging energy grids. It's thought the electrification of Africa won't follow the twentieth-century Western model of big power stations delivering centralised power, but a rather more diffuse mix of smaller solar, wind and geothermal plants. This requires distributed energy storage too and every town could have a gravity battery alongside a solar farm. The company's bridgehead is South Africa, where it is teaming up with local energy providers. The country has ambitious renewable energy plans, a legacy of coal mines with some shafts 3 kilometres deep, but a frail grid where blackouts in peak times are common.

'At Gravitricity we're working to enable the lights to be turned on in poorer regions where there currently is no grid. Africa can do things differently. Long-life, distributed electricity storage delivers the chance to build a grid that is designed from the bottom up for solar and wind, and it could be done at much less cost than replicating European or North American grid infrastructure.'

Desirable destination

Effective energy storage is vital to the full deployment of renewable electricity generation. This, alongside nuclear power, could eliminate fossil fuel power stations and the 25 per cent of all greenhouse gas emissions that they generate.

How to get there

Massive investment in all forms of energy storage: gravity batteries, liquid air, hydrogen, ammonia novel electric batteries and heat storage.

Fringe benefits

- Reliable electricity supply in the global south.
- High-tech high-skilled jobs in energy and related industries.

4
Blown Away

I like bats: their stealth, their mystery, their night 'vision' superpower and their faces like a cross between a lion and a werewolf. I like wind turbines too as they're emerging as one of the cheapest and most powerful tools for arresting climate change. It's a shame the two don't get along. A single turbine can kill hundreds of bats a year, a wind farm thousands. But bats can harm turbines' life chances too as humans trying to protect them can block proposed developments or insist they are switched off for long periods. Hubert Lagrange shares these two passions of mine but *he* has done something about it.

'Our technology has saved the lives of thousands of bats and, by keeping the turbines running, thousands of tonnes of carbon dioxide,' he says.

Hubert first encountered bats aged eight. His young ears could hear the ultrasonic clicks of larger species as they flew round his grandmother's barn near Dijon in France.

'They were so mysterious and as a kid, I wanted a closer look at everything. At dusk, when I could just see them,

I threw up insect-sized stones they would try to catch. I once even cast up fishing flies on the end of a line – no hook, of course.'

He went on to found Sens Of Life, an engineering consultancy dedicated to providing protection solutions for bats, birds and the wind industry.

Unfortunately, bats seem to be attracted to turbines. A recent study in the science journal *Nature* discovered pipistrelle bats – one of the most common – were 37 per cent more active around turbines than in nearby turbine-free locations. It's thought the giant windmills' slight warmth attracts insects and the bats follow. But how the bats die was something of a mystery. Turbine blade tips can move at over 300 km/hr, so collision seemed probable, yet bats found dead beneath them showed no evidence of impact. The truth is rather grisly: their lungs explode. A spinning turbine blade generates an envelope of very low pressure on the convex side, up to a metre thick. The bat flies in and, like a diver surfacing too rapidly, their blood vessels burst. Their ultrasound navigation might see the blade but is blind to the accompanying death zone.

Traditionally, the solution to this has been to forbid wind farms in areas with high bat populations or insist that they are switched off – 'curtailed' in the industry jargon – when the bats are most active. That is typically night-time with light winds and moderate temperatures and, added up across a year, can mean a 10 per cent power loss for a wind farm. A sizeable hit that potentially makes the difference between whether a wind farm is viable or not in some locations.

Hubert's company deploys ultrasound detectors to scan for bats and then combines that with weather data and knowledge of their behaviour. A computer program then assesses the risk and, if sufficiently high, will trigger the stoppage of the particular turbine by changing the pitch of the blades, slowing to a 'bat safe' speed in about 15 seconds. In day-to-day operations, artificial intelligence makes the decisions; no humans are involved. By turning off only when bats are actually in peril, Sens Of Life claim to be able to reduce production losses to below 1 per cent.

'We reduce bat mortality by 90 per cent. We could go higher but then you would produce less electricity. You need to find an equilibrium and ours seems fair.'

But it's not just the night-flyers in peril: birds are vulnerable too. The US Fish and Wildlife Service estimates there are between 140,000 and 500,000 bird deaths in America each year due to wind farms, though it should be noted that is much less than mortality due to buildings, cars and cats. Eagles and other raptors seem disproportionately affected, as much of their time on the wing is spent hunting animals on the ground: they are looking down and simply don't see the spinning blade ahead. It's not a threat they evolved to evade. In the States, wind farms can be fined millions of dollars or even shut down if they kill too many birds and traditionally they employ watchers to spot eagles and order a 'curtailment' if a bird-strike could be imminent.

But now cameras combined with artificial intelligence are providing a better solution for both occupants of our airspace. An American company, IdentiFlight, offers a network of cameras constantly scanning the wind farm for birds. They analyse the speed, position and trajectory of flight, and the software has 'learned' to identify different species. All this needs to happen while the bird is more than

500 metres away as they move so fast. When a valued species at risk is detected, the nearby turbine is automatically shut down. Early studies suggest that it protects more eagles and allows the blades to keep spinning more of the time.

Sens Of Life have deployed another sky-scanning system where wind power meets bird migration routes. Turbines in such flightpaths are routinely switched off for ten days twice a year, resulting in massive green energy losses. With accurate detection technology, Hubert Lagrange says this can be reduced to just a few hours' downtime per year.

So, this suite of hi-tech solutions will increase the energy yield from wind farms and reduce bird deaths, and it should help to reduce public opposition. These all matter hugely because wind power is growing so rapidly. We've already covered offshore wind in the 'Bladebug' section (see Chapter 1) but growth on land is equally staggering. The International Renewable Energy Agency, IRENA, reckon we will have around ten times more onshore wind energy by 2050. That would mean about 5,500 gigawatts of installed capacity, requiring an area a little bigger than Egypt. That assumes a 7 per cent growth rate every year, which may seem rapid but is actually far *less* than has been achieved on average in the last 20 years. Given that, in many countries, the cost of electricity produced by the wind is now lower than the market price, so it needs no subsidy, this prediction begins to look quite conservative.

In the last decade China has become the fastest wind developer, now overtaking Europe in terms of total onshore capacity, with America third. There is huge potential in the less developed markets of Asia, Africa and South America but even the global North is far from saturation. Just like offshore, size matters when it comes onshore turbines as the wind is stronger and more consistent higher up, enabling more power for the same land area. So, ageing smaller turbines are being replaced with bigger descendants.

One typical onshore wind turbine can supply enough electricity for 1,500 average European households; an offshore one can produce double that. They are our most visible weapon in the fight against climate change and some people don't like that. But with each swish of the blade cutting carbon emissions, they are preserving our world, not spoiling it.

Desirable destination

5,500 gigawatts of onshore wind generation capacity by 2050, 10 times more than today and enough to cut around 15 per cent of our current greenhouse gas emissions

How to get there

- Increasing turbine size.
- Greater public acceptance through familiarity and, where appropriate, more local ownership or benefit.
- Further reductions in conflict with wildlife.
- Increased electricity storage capacity on the grid to accommodate more variable renewables.

Fringe benefits

- Greater protection for wildlife.
- Improved energy security for countries with no fossil fuel resources.

5

Scalding Solar

If ever a man was destined to be a creative engineer it is Faisal Ghani.

'It's incredibly cool to think my job involves using energy from a star 93 million miles away. I still get a buzz out of that – I guess it shows what a nerd I am. Aged 12, I made a space probe at school in Sydney, complete with launch rocket and solar panels. But a teacher moved it and it went off in the classroom.'

He's now a solar pioneer working in the Scottish city of Dundee and clearly possesses a sense of both irony and optimism: 'If my inventions work here then we're sorted for much of the world.'

We meet in his warehouse crowded with flatpack boxes stacked on pallets about to be shipped to Africa. Each box is about half a metre square and weighs just 10 kilos. The contents can be assembled in minutes to make a square-based pyramid with glass sides and a black pipe coiled in a cone shape within. The pyramid turns sunshine into hot water. This chapter is about heat, not electricity.

'Under an African sun a single unit can heat 75 litres of water to 50°C in a day.'

Turning sunshine into heat – so-called solar thermal – is far more efficient than converting it into electricity. Photovoltaic panels today use only about 20 per cent of the energy that falls on them whereas more than 50 per cent can be turned into heat: Faisal Ghani's solar pyramid, the SolarisKit, does it simply, cheaply and with surprising beauty. In his own words: 'We have engineered the engineering out of it. The first version had 14 bolts and nuts but this one is just clip together. You can build it in 20 minutes. We hope to sell them for around £100 each and they are compatible with existing water tanks.'

The technology has had keen interest from campsites and holiday parks in Europe but Faisal and his team are really interested in deployment in poorer countries, especially in Africa. Carbon emissions are growing fastest in the global south and there is desperate need for practical low-carbon solutions. The SolarisKit team have partners in Rwanda where research suggests that just over a quarter of household income goes on water heating for bathing and washing. Solar thermal is not entirely new to Africa – you can see tubes and tanks on the roofs of quite basic housing in South Africa – but Faisal Ghani says most of these systems have been developed for Europe and are still quite complex and pricey: 'Affordability and simplicity are the key to deployment, especially in the south. But the media and even fellow engineers can be snobby about simple solutions to climate change. They think artificial intelligence, big data and "the internet of things" are "sexier". I fundamentally believe solutions like ours will make the most impact.'

The International Renewable Energy Agency says solar has huge growth potential and, although Faisal Ghani's hot-water pyramid is a piece of home-scale design genius, solar thermal can be remarkably high tech.

The Belgian company Azteq is reaping the heat from Flanders's fields as well as parking lots and dockside gantries. Its technology resembles a photovoltaic array crossed with a skateboarder's half-pipe. The array is in fact made up of parabolic mirrors 6 metres across and typically 120 metres

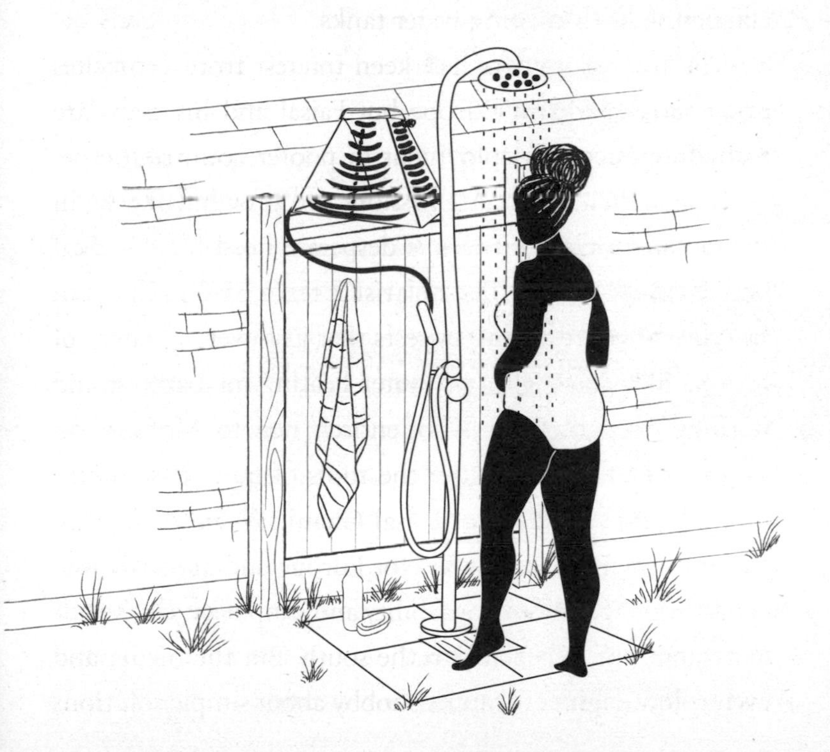

long. The parabola reflects sunlight on to the focal point of the tube where there is a pipe filled with a special oil designed to absorb and carry heat. Each array of mirrors pivots to track east–west following the sun. The result is a supply of heat reaching 400°C (752°F). Simpler versions of this concentrated solar thermal technology have been around for some years but largely confined to desert locations with guaranteed daily sun. By improving efficiency and coupling it with thermal storage technology, Azteq has made it viable for more variable northern climates.

Co-founder Peter Vandeurzen sees so many applications, especially in the European Union, which has set a carbon emission reduction goal of 55 per cent from 1990 levels by 2030. 'Many big users of heat find switching to electricity is not feasible so greening the power sector won't help. We provide a low-carbon switch for chemical companies, breweries, food manufacturers and district heat systems who traditionally use gas boilers. We had a cold snap the other day here in Belgium, still sunny but down to -15°C (5°F) and the tubes were producing heat like hell.'

Just like solar electric panels, they do need space but they are light enough to be mounted on rooftops or supported on gantries. In the port of Antwerp, two ranks of mirrors stand proudly above the dockside. Cars, trucks and even train wagons move unhindered beneath. And in case you are worried about a passing dove scorching their feet on a tempting but 400°C perching rail, the fluid runs in a double-skinned pipe wrapped by a vacuum.

In the nest of renewable technologies, solar thermal's growth has been rather stunted in recent decades as its electrical sister gets fed with abundant investment and invention. But as we come to realise that a climate-friendly economy can't fly without decarbonising heat, Peter Vandeurzen thinks all that will change. His earlier professional life was spent in the digital sector and he sees parallels: 'It's like the internet 25 years ago. We are standing in front of the next big change. And just like that time, I'm having too much fun!'

In a small way, I've been on this journey myself. Ten years ago, I removed the oil-fired heating from a house in the Western Isles of Scotland and to warm the water I installed a solar thermal system on the roof and a bigger tank inside. It heats the water and anything extra goes to a couple of radiators. The results have been stunning: it was about half the price of setting up photovoltaic panels and delivers more than twice the energy. OK, I can only use it for heat but that's pretty crucial in the Hebrides – and worldwide.

Desirable destination

4% reduction in greenhouse gas emissions through widespread uptake of solar thermal as a domestic and commercial heat source.

How to get there

Education: ending solar thermal's 'Cinderella' status.

Regulation: rules to restrict high-carbon heating systems and incentivise renewables.

Invention: engineering season-long heat storage tanks and ways of integrating solar thermal with hotter technologies.

Fringe benefits

Less deforestation and air pollution from wood burning in poorer countries.

6
The Nuclear Option

'I've gone from making baldness reversal products to a climate change reversal product.' This is the journey of Ian Scott, once chief scientist for the consumer goods giant Unilever, now the driving force behind a new type of nuclear reactor. Moltex's stable salt reactor can run on nuclear waste, the company claim you could picnic in sight of its exposed core and they aim for it to be a cheaper source of electricity than gas.

Let's first take a step back and address the radioactive elephant in the room: does a zero-carbon world need nuclear power at all? The technology is perceived by many as dangerous, appears to be ever-more expensive and leaves toxic waste for generations. Unfortunately, though, our cheapest and most rapidly growing renewable energy sources – solar and wind – are as uncontrollable as the weather. Geothermal or tidal – which are constant or you can set your watch by respectively – are limited, underdeveloped and currently expensive. Energy storage is emerging, as discussed earlier in the book, but still lacks the capacity for the volume of raw energy required to keep us powered up for long periods. A viable, acceptable and cost-competitive nuclear energy source makes a zero-carbon future much more plausible.

Moltex has been chosen to build a new nuclear power station, fuelled by a molten salt reactor, in the eastern Canadian province of New Brunswick. The final cost will be around US$1.2 billion and it should be completed by 2032. Ian Scott passionately believes it will be a game changer: 'There is a very, very serious risk that something will go very wrong with the climate. If we do not get nuclear power providing a big chunk of our energy, we are in serious trouble.'

Ian grew up on the industrial northeast coast of England where his father worked in the steel business. Winning a scholarship to the University of Cambridge, he spent his first year studying particle physics before he discovered his maths was too weak and switched to biosciences. His career in cosmetics and care products thrived before he retired in the early noughties.

'I was very happy in retirement but then I saw that the UK government was promising to pay new nuclear generators more per unit of electricity than I was paying in my domestic bill.' A cue for most of us to give a weary shrug, but not Ian: 'It drew me into looking at nuclear again as we are never going to make a difference to climate change if nuclear remains so expensive.'

Now approaching 70 years old, Ian Scott is championing a new type of reactor for power generation and, to understand it, there is no escaping a little nuclear physics.

All nuclear power stations use the intense heat from an atomic reaction to create steam to drive a spinning turbine to generate electricity. The most common are pressurised water reactors (PWRs). An advanced type of PWR is at the core of new power stations being built in the UK. It is worth pointing out that, measured in terms of deaths per unit of energy produced, nuclear is by far the safest electricity source. It has a mortality rate 1000th that of coal and 300 times lower than oil. But it's the fear of what *could* happen that drives judgements on nuclear.

In a PWR, the atomic reaction occurs within ceramic pellets of uranium generating huge amounts of heat and pressurised gas. These are held within metal tubes surrounded by water, which itself gets very hot: 300 to 350°C (572–662°F). In normal conditions, of course, water boils at 100°C (212°F) so, to avoid that, it is held under enormous pressure in steel and concrete chambers. It's that pressure that means PWRs are fundamentally hazardous and only made safe by expensive containment and cooling, whereas, according to Ian Scott, molten salt reactors are intrinsically safe. So safe in fact that they were first suggested back in the 1950s for nuclear-powered planes as aircraft might crash and you can't have an accident that wipes out a city.

In a molten salt reactor, the radioactive elements such as uranium or thorium are held in a liquid sodium chloride solution. Molten salt also cools the core and transfers the heat into steam to drive turbines. Its great advantage is that it remains liquid at high temperatures without pressurised

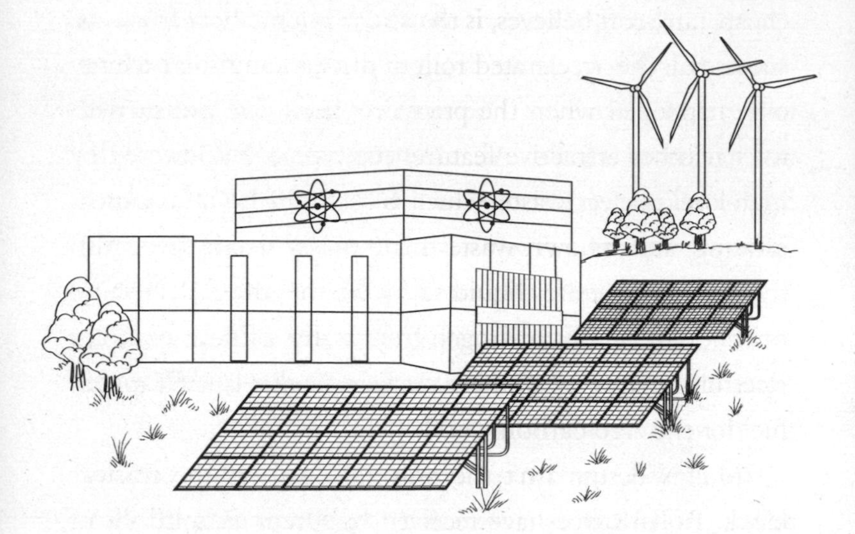

containment and creates far less radioactive gas. Stand close up and you could still get a lethal dose of radiation but it's never going to wipe out a city or contaminate a continent.

'Let's try this as a movie analogy,' suggests Ian. 'The PWR is like the super T. rex in *Jurassic Park*: only made safe by a strong cage and woe betide the neighbourhood if the containment fails. Our molten salt reactor is like a T. rex with no legs: you don't want to want to walk up next to it but it's never going to burst out and threaten anyone. If the power failed to our reactor and all the staff walked out, it would just gradually cool down and turn off. It is safe by design.'

Nuclear safety is obviously good but the molten salt reactor could also be cheap. It's all the super-strength engineering and multiple fail-safe mechanisms that make conventional nuclear reactors so expensive and big. Being cheap, Ian Scott believes, is the only way that nuclear can be successful: the accelerated rollout of wind and solar energy only happened when the price dropped. The Moltex reactor has other attractive features too: it can be powered by high-level nuclear waste, which is currently housed at great expense, and its own waste is much less radioactive. Also the reactor generates liquid salt at above 600°C (1,112°F) – hot enough to split hydrogen from water without complex electrolysis, and, as we have seen, hydrogen is itself a vital fuel for the zero-carbon world.

Moltex design isn't the only new kid on the nuclear block. Rolls Royce have received hundreds of millions of pounds from the UK government to develop small modular reactors which they believe can be cheap thanks to the economies of building on a production line system.

So why do these promising new ideas take so long to realise? Molten salt reactors do come with their own hazards around potential corrosion of critical components, using waste fuel is challenging and much of the required control system engineering is unproven in a nuclear power station setting. But Ian Scott believes there is also a mindset hurdle. For all its superficial sophistication and modernity, nuclear energy is rather innovation-averse. I've been around some nuclear power stations and the dials, switchgear and control rooms make you feel like you've stepped on to some vintage Hollywood sci-fi set. The PWR is a multi-billion pound embodiment of the old adage 'Better the devil you know'. New PWR designs have vastly improved safety margins but

Ian Scott questions whether such caged threats should be getting regulatory approval at all when intrinsically safer designs are available. After a career in consumer goods, he recognises he is a stranger in the atomic world. 'Bringing an outside perspective of what you can do quickly and innovatively is a huge strength. I think the attitude in the wider nuclear industry is "Thank God somebody is trying to do this but you've got no idea how hard it's going to be."'

The Moltex team know that regulators and the public will remain sceptical until their molten salt reactor has a safe operational track record. But given his determination to deliver cheap nuclear power, Ian Scott might well do it.

Desirable destination

By 2050, replace all coal-fired power stations, which currently produce about 22 per cent of greenhouse gas emissions. This would require more than tripling present nuclear power capacity.

How to get there

Technology and policy: novel types of nuclear reactor such as molten salt will need a track record of safety, value for money and approval by regulators.

Compatibility: ensure nuclear power works with other low-carbon solutions such as energy storage, hydrogen production and combined heat and power.

Security: make sure nuclear materials and plants are protected from hostile forces.

Attitude: persuade some environmentalists to embrace atomic energy, not demonise it.

Fringe benefits

- Disposal of existing nuclear waste.
- Less air pollution from coal burning.

7

Helpful Hydrogen

Hydrogen is the most abundant element in the Universe and could save our little speck of it. Hydrogen works in a similar way to fossil fuels in that it combines with oxygen to create energy, but it lacks their fatal flaw: the reaction emits no greenhouse gas. There is no carbon in the formula, just hydrogen plus oxygen-yielding H_2O – water. Right now, in the early 2020s, companies, governments and investors are flooding towards hydrogen.

But, if hydrogen is so great, how come it has taken so long for us to realise the attraction? The answer is two parts scarcity, one part history. Though H_2, the simplest of molecules, makes up 70 per cent of the known Universe, here on Earth it's almost non-existent in its pure state. It is very chemically attractive, readily bonding with many other elements, especially oxygen to become water. Hydrogen gas is also so light, just 7 per cent the density of air, that if not contained it vanishes upwards. This levity lifted the giant airships of the early twentieth century but its volatility ensured their demise: both the British *R101* and the

German *Hindenburg* perished along with many lives in a ball of flame as they crashed to the ground. The latter was caught on film, embedding an enduring formula in the popular imagination: hydrogen + humanity = danger.

It was the technology behind another form of flight that began hydrogen's rehabilitation – space flight. Fuel cells powered NASA's space capsules from the 1960s and their key ingredient is hydrogen. A fuel cell generates electricity through an electrochemical reaction, not combustion: hydrogen and oxygen are combined to generate electricity, heat and water.

Ceres Power, a UK-based company spun out of London's Imperial College 20 years ago, is now at the forefront of developing and deploying fuel cells for homes, businesses and transport.

'It is an idea whose time has come,' says chief executive Phil Caldwell. 'Many people used to think we could save the world with renewable electricity – and it is incredibly important – but still around three quarters of our energy

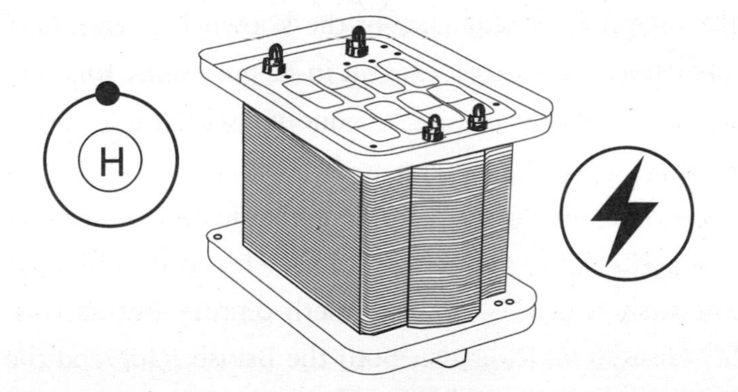

comes from fossil fuels. That's a chemical energy supply, but we can decarbonise it by using hydrogen instead.'

Even given the rapid growth of renewable electricity, there are many tasks that still depend on chemical fuel, a power source based on molecules not electrons: industrial manufacturing, steelmaking, cement plants, fertiliser factories, heavy transport and aviation. Hydrogen is the climate-friendly solution and the world has realised.

'If you had a webcam on the reception of our local hotel, you'd see all the nationalities coming to see what hydrogen can do. Ten years ago, it felt like we were pushing this technology, now customers are pulling it.'

Ceres's product – the Steel Cell – is now manufactured under licence in Germany and deals have been signed for production in South Korea and China. It is a multilayer sandwich of steel and ceramics that has been refined through interaction with customers. Mark Selby, Ceres's chief technologist, believes this two-way street between invention and application is critical, unavoidable and creates value: 'No new technology arrives straight from the inventor as a low-cost, robust, popular product. It always needs refinement through real-world use and efficiency through scale of production. That's what happened with solar panels and batteries; it's now happening with fuel cells.'

They are, in effect, mini power stations producing electricity and heat in a growing number of offices, residential blocks and businesses. Data centres, the digital brains of our modern world, are among the keenest customers. Many of the big tech companies are very keen to drive down their energy footprint and their biggest fear is power cuts: a fuel cell powered by the gas grid ticks their boxes.

'Hang on,' you are thinking, 'did I read *gas* grid? What is so green about that?'

That is the cunning thing about fuel cells: they work with the mains gas of today and they will work with the hydrogen of tomorrow. Supplied with natural gas, fuel cells produce 30–50 per cent less carbon dioxide than conventional natural gas boilers and mains electricity, and the very same fuel cell will work with zero-carbon hydrogen. They are, in the parlance, 'hydrogen ready'.

But as pure hydrogen doesn't occur on Earth, we'll have to make it, and fuel-cell technology works here too. Essentially, you reverse the equation and use electricity to split water – H_2O – into hydrogen and water. As long as your electricity source is zero carbon, such as wind and solar, the resulting hydrogen won't contribute to global warming. Denmark is planning a huge offshore wind electrolysis plant designed to supply a quarter of a million tonnes of hydrogen fuel a year by 2030 for use in buses, trucks and planes. Germany has an electrolyser target nearly five times larger for the same date. Even the oil-rich Gulf states are now thinking of using another natural resource they are blessed with – sunshine – to power large-scale solar hydrogen production.

It is easier to acquire hydrogen by splitting it off the methane molecule (CH_4), otherwise known as natural gas, rather than water, and this is how most hydrogen is made today as it's a core process in chemical fertiliser plants and has been for a hundred years. But methane is a hydro*carbon*, so this process leaves you with carbon dioxide exhaust. If this is simply released into the atmosphere, as happens in nearly all fertiliser factories, the hydrogen produced lacks any climate merit. But, if combined with the emerging technology of carbon capture and storage, this could be a high-volume hydrogen source appealing to countries such as the USA, Norway and the UK that have established infrastructure and jobs in the natural gas sector.

There are financial costs and efficiency losses in turning electricity to hydrogen and back again, but these are steadily shrinking and many observers think we are approaching a world of electro-chemistry where energy is regularly transformed from one form to another according to demand.

Another approach would be to pipe the hydrogen directly to the customer. If my central heating boiler ran on hydrogen it would be zero carbon. Sadly, the gas grid isn't built to cope with pure H_2: it weakens steel pipes and some couplings won't stop leaks of such a small molecule. But you can mix it with natural gas. The British clean fuel company ITM Power reckons you can add 20 per cent of hydrogen into the network without leakage and without any need to replace cookers or boilers.

So, what about those airships? Is hydrogen safe? Compared to natural gas or petrol we have around us much of the time – yes. At 14 times lighter than air it rapidly disappears upwards rather than pooling on the ground like liquid fuel or even gas, which gathers on the floor before catching a spark and blowing your house apart. Also, hydrogen's explosive power is far less than familiar fossil fuels. Our suspicion of hydrogen is simply fear of the unknown stoked by one searing historic image. We need to get over it: hydrogen is the fuel of the future.

Desirable destination

Switch heavy road transport – trucks and buses – to hydrogen by 2050, saving 6 per cent of our emissions. Using hydrogen to steadily replace fossil fuel in the gas grid and for industrial uses could save at least another 10 per cent. Creating hydrogen from solar and wind will be vital for renewable energy storage.

How to get there

- Complete redesign and rebuild of fossil fuel engines and infrastructure.
- Public tolerance of higher energy prices.
- Massive investment in clean hydrogen generation through electrolysis, and the carbon capture and storage required after separating hydrogen from natural gas.

Fringe benefits

- Cleaner air.
- Huge employment potential in the energy industry and related industries.

8
Solar Flare

Solar power is now the cheapest source of electricity in history. According to the International Energy Agency, it can now be produced for US$20 per megawatt hour. That means if you plugged in the average UK house and were only charged the wholesale price, your annual electricity bill would be about US$30. Globally between 2010 and 2020 solar generating capacity grew 30-fold to 600 gigawatts – more than half of the USA's electricity demand. On 30 May 2020, the sun delivered one third of electricity demand in the UK, a country famous for umbrellas. Solar is the overachieving child of renewable energy generation but its school report still reads 'could do better'.

Solar's success has come from the plummeting cost of making the panels. The fuel – sunshine – is free, so the price of solar electricity depends on how cheaply you can make and install the equipment. The first solar – technically photovoltaic or PV – cell was created in 1883 but practical uses were scarce until they found favour in the space race for powering satellites. The oil crises of the 1970s and 1980s, when war in the Middle East drove up fossil fuel prices, spurred further interest. By the early twenty-first century, growing demand for renewable energy drove manufacture by big electronics companies, such as Siemens and Panasonic. But it has been the power and scale of the Chinese solar industry that have really slashed the price and boosted installation.

What hasn't changed so much in recent years is the technology of the cells themselves or their performance. The core ingredient is a layer of silicon, a semiconductor with a crystalline molecular structure, which converts the beam of photons from sunlight into a flow of electrons delivering current and voltage. For most panels deployed around the world today, only about 15–20 per cent of the sun's energy hitting them is actually made into electricity, a figure that hasn't budged much in the last 20 years.

Henry Snaith and his team at Oxford PV can do better: 'Over the next decade we will see a massive growth in solar deployment and what we are pushing desperately to do is have our technology ride and enable that wave. We have to deliver PV electricity at an economically viable rate all over the world, not just in the sunniest places.'

Snaith is a physics professor and serial winner of scientific awards. He, and now the company he founded, are dedicated to improving the efficiency of solar cells. It's all about the spectrum. Sunlight is made up of different wavelengths of light revealed in the varied colours of the rainbow: red, orange, yellow, green, blue, indigo, violet. Silicon only reacts to the red end of the spectrum and energy from rest of the sunlight is largely lost as heat. Snaith's idea is to combine silicon with another material that can harness energy from the blue end of the spectrum and then you can get more clean, green electricity from each panel. So, what is the newcomer grabbing the sunlight in this high-wattage double act? Perovskite.

Perovskite is a naturally occurring mineral – calcium titanium oxide. It was originally found in the Ural Mountains and named in honour of a Russian mineralogist, Lev Perovski. It is also a semiconductor with a crystalline chemical structure that can be made in the lab with much less energy than that required to make panel grade silicon. Its potential in solar panels has been known for a few years and a handful of companies is racing to commercialise production. Oxford PV's tandem cell, the perovskite silicon sandwich, has already achieved 29 per cent efficiency, which is one and a half times better than regular panels. The theoretical maximum is 45 per cent.

'These perovskites have just emerged in the scheme of things. They've just come out of the lab and we're trying to drive this forward to manufacturing on a very short timeline. Most of the rest of the industry think perovskites are coming in five to ten years. We are showing that it is happening now.'

Oxford PV is due to start commercial production in a factory in Germany in early 2022. Snaith acknowledges that initially his panels are more expensive than simple silicon competitors as they don't yet have the economies of scale,

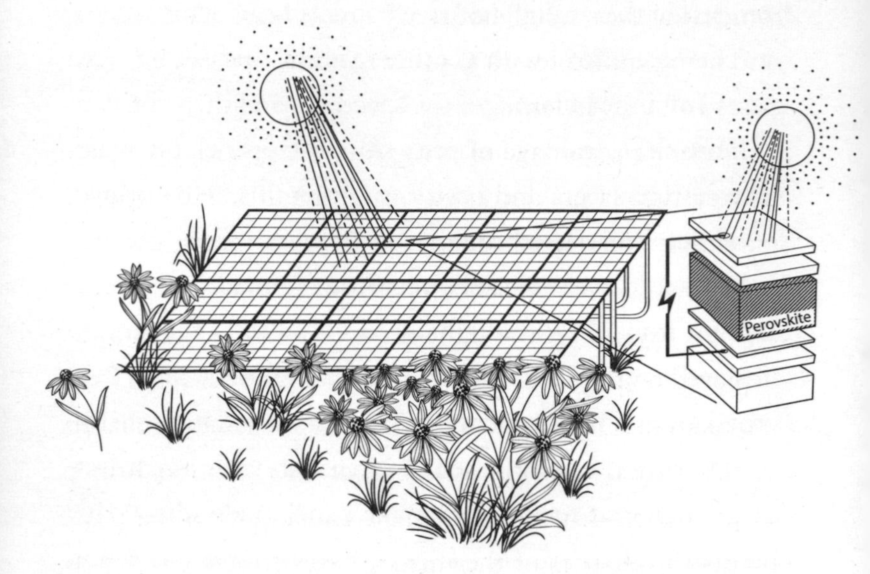

and he doesn't expect to see them covering large fields on solar farms soon. But on settings such as rooftops, where space is limited and installation costs can be steep, the 'higher price/higher performance' model works well as it can deliver a swifter payback on investment.

The other question mark over perovskite cells is longevity. Silicon solar panels have proved remarkably robust with performance typically dropping by less than 20 per cent even after 25 years of use. Snaith has known from the start that making perovskite panels stable is crucial and says his company's focus has been on ease of manufacture and stability in use alongside great technical performance. Until they can prove durability though time in the real world, new products must rely on certification from the International Electrotechnical Commission based in Switzerland, which involves 1,000 hours of exposure at 85°C (185°F) and being chilled to -40°C (40°F). Oxford PV's panels have passed all tests so far.

Another advantage of perovskite is that it can be made in very thin layers and used on flexible materials or even glass that still lets through the light. Window coatings would have lower efficiencies of around 10 per cent – after all, you have to let the light through – but they could be deployed on the sheer shiny face of a skyscraper. That's not the focus for Oxford PV and Snaith welcomes more players in the perovskite world as being totally exceptional often makes investors nervous, whereas a small perovskite posse boosts the chances of the material accelerating an already booming industry.

Snaith says, 'It is mission critical. At the moment, we produce just a few per cent of our overall energy needs from solar. We are on the verge of a transition to vastly improved technology to deliver vastly lower carbon emissions.'

How vast? Well, solar energy is already projected to grow 14-fold by 2050 and should deliver a 12 per cent cut in our total carbon emissions using existing silicon cells. If these were all replaced by perovskite sandwich cells that figure could rise to 18 per cent, and if panels get more efficient again as Snaith is predicting then … who knows? But the future is bright.

Desirable destination

Elimination of 18–24 per cent of our greenhouse gas emissions by 2050 through deployment of more efficient solar panels.

How to get there

Research and development: increased investment in perovskite panels and further innovations to squeeze more wattage from the sun.

Deployment: more insistence from planning authorities that new buildings should have solar roofs.

Affordability: continued decline in the price of solar panels.

Fringe benefits

- Cleaner air.
- Creation of hi-tech job opportunities.
- Affordable electricity in the global south.

Nature

9

Sublime Seagrass

It's mid-October and the wind has swung round to the north and into the mouth of the bay. This clears the air but clouds the sea as sharp waves churn the silt. It's the water I'm peering through, trying to spot grass beneath my fins, grass growing on the seafloor. More accurately, I'm trying to spot bare patches on this underwater lawn as I am among a group of scientists sowing seagrass. We are all bobbing about in wetsuits beside Porthdinllaen, a craggy outcrop looking like a barb on the fishhook-shaped Llŷn Peninsula in north-west Wales, home to the country's largest subsea meadow.

On a bright day, beneath clear water, seagrass looks like a lush summer meadow: vivid green blades around 30 cm long gently swaying in the passing current. Like grass on terra firma, seagrass relies on photosynthesis for growth; it has flowers, pollen and pollinators but 'down where it's wetter', the pollinators are crustacea not insects. It's also a huge, and potentially growing, carbon store. Per hectare, some seagrass beds are thought to be 35 times more rapid at gobbling greenhouse gases than a tropical forest. It covers just one five hundredth of the seafloor but is thought to absorb 10 per cent of the oceans' carbon. A chilly dip on a blustery day seems like a small price to pay for enlarging such a critical carbon sponge.

Seagrass likes to grow on sand or silt at depths of around 1–5 metres but can grow much deeper where clearer water and bright sun allow sufficient light to reach the leaves. The plant used to form a widespread green fringe to bays and estuaries but, in the northern hemisphere at least, it's been vastly diminished by pollution, dredging and recreation. The UK, for instance, is thought to have lost 90 per cent of its seagrass. The World Resources Institute estimates that conserving and restoring seagrass beds is one of the most effective ocean-based climate change solutions.

Richard Unsworth is chest deep, paddling about and occasionally ducking down to inspect the seafloor. He's from the Seagrass Project, a charity based at Swansea and Cardiff universities in Wales, and floating beside him is an inflatable table carrying bags of seed. About 50 seeds mixed with sand are held in each mouse-sized hessian sack. They have been painstakingly collected by plucking some of the long lime-green seedpods where seagrass is abundant. On finding a good spot, Richard takes a bag and fixes it to the bottom with a small wooden stake: dedicated work he likens to allotment gardening. The scourge of seagrass here is dragging boat anchors ploughing a furrow through the meadow or scouring from swinging mooring chains. This damage needs to be repaired.

At another restoration project in Wales, his team have taken a more mechanised approach, running a rope, with seed sacks attached every metre, off the back of a boat in Pembrokeshire's Dale Bay. The rope and the bags rot away, leaving the seed to grow. It's working: Richard Unsworth's 'seeds of hope' have delivered. 'We've seen our green shoots of recovery reaching up to the sunlight. It's hard confronting the climate and nature crisis, especially as a father with young kids. But it helps me and them to understand it through this restoration work.'

The power of seagrass to capture carbon so rapidly stems from the way it grows and its effect on the surrounding seawater. As it grows through photosynthesis, carbon is absorbed from the water into the stalk and leaves of the plant. Growth is unusually rapid with a high turnover rate so these leaves frequently break or die and fall to the floor, where they build up like leaf litter in a forest. But seagrass has another talent, unmatched on land: it can trap carbon-rich scraps by slowing the movement of water. Vigorous currents are impeded by friction on encountering the lush meadow and this means sediment suspended in the water drops down. Flakes of seaweed, seagull droppings, micro-plankton – anything floating in the water column – is more likely to fall in a more leisurely current. It's the same physics that means rivers silt up as they meander gently through lowlands. This becomes combined with the roots of the seagrass to form a deep, soil-like layer that permeates down into the seabed. In the Mediterranean, where seagrass grows

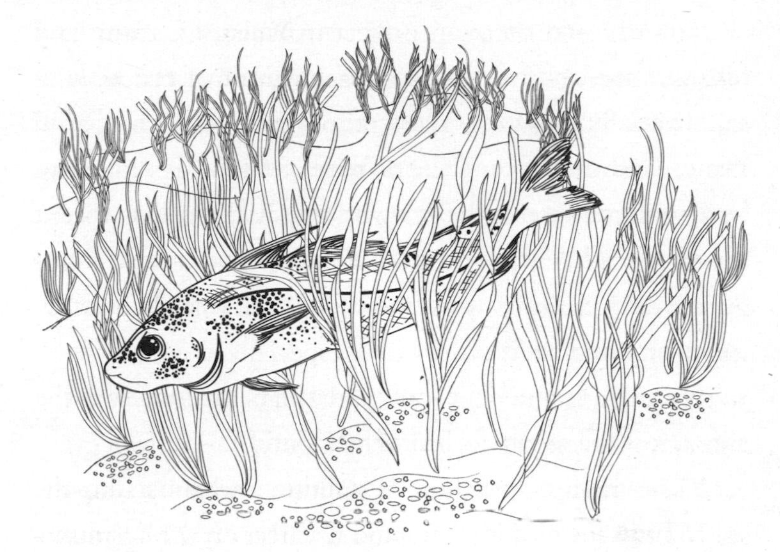

exceptionally well, these carbon-rich reserves can be tens of metres thick. On average, seagrass beds are thought to store twice as much carbon per unit area as a forest: 83,000 tonnes per km^2 as opposed to 30,000 tonnes per km^2 for woodland.

By far the biggest restoration project so far has been in the Virginian lagoons on the USA's Atlantic coast. In the 1930s, the combined effects of a disease and a hurricane eradicated the seagrass from much of the US Eastern Seaboard, including these lagoons, and with it went a successful scallop fishery. By the late 1990s, studies suggested that the only thing holding back the return of seagrass was the absence of plants to spread seed. So, 74 million seeds were harvested elsewhere and sown in various plots across 2 km^2. Now, 20 years after the project began, the natural spread of plants and dispersal of seeds have extended the seagrass cover to over 30 km^2 and there are so many benefits. Carbon and nitrogen are being stored in the sediment of the beds at exponentially growing rates, the water is cleaner and clearer, and then there's the wildlife. Seagrass is known to be a vital and vibrant habitat. In Virginia, that has meant the return of bay scallops, silver perch and pinfish. In European waters, seagrass is cherished as a nursery for juvenile cod, plaice and haddock.

But sadly, seagrass meadows are still being lost, with the area across the world declining by around 1.5 per cent every year. The main culprit here is pollution running into the sea from industry and agriculture, particularly phosphates and nitrates from farming and detergents, which result in what's called 'eutrophication' – an excess of nutrients in the water that kills the seagrass. In many parts of the world, authorities are trying to address this pollution and there has been some improvement in the European Union thanks to tougher rules for industry, farmers and water companies. This has prompted restoration projects in many places including the Venetian lagoon and the Balearic Islands in the Mediterranean to the waters around Scandinavia and the Baltic Sea. In British waters there are plans to improve or return seagrass beyond Wales to Plymouth, Southampton and spots along the east coast from Essex to the Firth of Forth in Scotland.

Desirable destination

Absorption of an extra 1–2.5 per cent of our emissions by 2050 from the reestablishment and expansion of seagrass, salt marsh and mangrove ecosystems.

How to get there

- Preservation and restoration of thousands of square kilometres of seagrass through tighter regulation of pollution and coastal development alongside planting programmes.
- Protection of mangroves.
- Reestablishment of salt marshes.

Fringe benefits

- Better habitat for fish and other marine creatures.
- Greater abundance of stocks for commercial fishing.
- Natural storm and flood protection for coastal communities.

10

A Mammoth Task

Welcome to a place where trees are the problem, not the solution; to a place where flatulent ruminants should be bred not blamed; to a place where our climate-friendly presumptions come to die. Welcome to the Arctic.

A pioneering family of Russian scientists, backed by credible academics worldwide, is pushing a theory that sounds, at first, like sacrilege: it would slow global warming enormously if the Arctic lands looked like a chilly African savannah with broad grasslands, a few trees and herds of grazing animals. This, they say, was the natural state of the north before mankind hunted the mammoth to extinction and pushed most other big herbivores to the edge. Their return en masse would help save the world.

Nikita Zimov runs the Pleistocene Park. Founded by his father, Sergey, it is 144 km^2 of eastern Siberia grazed by reindeer, musk ox, yaks, moose, horses and bison. He would love to have mammoths too but their rebirth remains science fiction for now. 'There are almost no trees naturally in the Arctic and we should go back to that. What people now see as natural is almost 100 per cent invasive species. Humans massively changed the planet 15,000 years ago, causing mass extinction. The trees marched north.'

To see how their removal could help with climate change, you need to get a handle on the thermodynamics of the Arctic.

The ground across vast swathes of North America, Scandinavia and Russia above 60 degrees north is frozen. This permafrost can reach down for hundreds of metres and it keeps the land in an icy stasis. But temperature rise due to human-made climate change in the Arctic is around 2°C above the pre-industrial level and twice the global average. The ice is losing its grip on soils that hold deep stores of peat and organic material. When that happens, just like a power cut in your freezer, microbes get active, the rot sets in and gases escape. Emissions from a widespread thaw would add around 4.5 billion tonnes of carbon to the atmosphere every year, roughly the equivalent of the world burning 50 per cent more fossil fuels, and would trigger a vicious cycle of climate change. It is a destiny to be avoided at all costs.

The problem with trees is their colour. Green in leaf and brown in winter, these dark colours absorb more of the sun's heat than white snow. Outside the winter months of very low sun, the Arctic is quite bright, especially in the spring and autumn when snow-covered ground reflects much of the sun's energy back into space. But trees trap more of that energy as warmth close to the ground. The result in the Arctic is milder air and melting permafrost.

Even when there is no snow, lighter grassland reflects more heat than trees. If you park a black car in the sunshine it warms up quicker than a white one.

Scientists call this the 'albedo effect' and there is a similar concern that the dark waters of the melted Arctic Ocean trap more heat than the disappearing white sea ice. Marc Macias-Fauria, biogeoscientist at the University of Oxford, says a forested landscape in the north absorbs much more energy than the tundra – in some months twice as much. 'In large regions of the north, the addition of trees contributes to global warming even when we account for the carbon locked up in the wood.'

More herbivores would reduce forest cover by stripping bark, nibbling leaves and grazing saplings but all those hooves also have another beneficial effect – trampling snow. The white stuff reflects the sun but it also insulates the ground. In the Arctic winter, the air temperature can plummet to -50°C (-58°F) and stable permafrost needs as much cold as possible to penetrate down. But a thick layer of snow keeps away the deepest freeze. Mountaineers dig snow holes to escape the worst weather and warm-blooded rodents can exist beneath a good covering: it literally is a *blanket* of snow. But grazers such as reindeer, Siberian horses and musk ox squash the snow underfoot and push it aside to reach the vegetation: transforming a thick white duvet into a threadbare sheet and the ground cools more thoroughly as a result. From work in the Pleistocene Park, Nikita Zimov and colleagues have shown that average soil temperature is

more than 2°C lower where you have grazing animals versus where you do not. In the winter, 25 cm below the surface, it is 15°C cooler. Marc Macias-Fauria says, 'In the coldest part of the winter there is a much deeper freeze when you have a bunch of grazing animals trampling the snow. This could help preserve the permafrost for decades.'

But surely all the carbon locked in the wood of trees counts for something? 'No, you are absolutely wrong,' says Nikita Zimov, stripping away another preconception and the Arctic trees' last line of defence. When you include roots and old foliage held in soil, grazed tundra stores more carbon than forest and it's less vulnerable to fires.

More problematic for herbivores' claim to climate friendliness are their methane emissions. Reindeer, bison and musk ox are ruminants. So, like cattle and sheep, their digestion process results in large quantities of this potent greenhouse gas coming from both front and back ends. This dents but doesn't undermine their beneficial effects in limiting climate change but does provide another motive for the teams in Russia, the USA, China and South Korea all trying to bring back the mammoth from the dead. Mammoths, like elephants, are not ruminants and so much less gassy.

According to Marc Macias-Fauria the mammoth steppe once stretched from Iberia to Siberia and across North America: 'It was the biggest single habitat the world has ever seen.'

Recreating a significant part of it would be a huge challenge for governments, ecologists and the people who live and work in the north. But it has enormous potential to arrest the perilous loss of permafrost and the rapid warming of the Arctic. A mammoth task.

Desirable destination

Preventing 2 per cent of future emissions by returning grazing animals to 20 per cent of the permafrost area. That is around 4 million km^2 or a little less than half the size of Canada.

How to get there

- Intensive breeding and reintroduction programmes of horses, reindeer, musk ox and bison – bison are especially important, in the absence of mammoth, for their ability to topple trees and kill them by eating the bark.
- Revive the mammoth with cloning, implanting frozen mammoth sperm into an elephant egg or gene-editing elephant DNA.

Fringe benefits

- Mammoths!
- Allow wolves to spread due to more prey species.
- An Arctic animal ecosystem as rich as the African savannah.

11
Good Logging

For just about anyone with a scrap of concern about climate change, the felling of tropical forest is close to the top of the 'don't do' list. But not for Peter Ellis, global director of climate science for the respected American environmental charity, The Nature Conservancy: 'Wood is good and the only way to get it is through logging. It is really important that loggers become our allies in the fight against climate change and protecting our environment.'

Now this might sound like recruiting the devil to fight damnation or using water to stay dry but stick with me. Logging for timber in the tropics already happens across a wooded area covering about 4 million km^2 – around the size of India. Now the fact that this amount of tree cover isn't *lost* every year reveals that most tropical logging extracts valuable trees from within the forest, leaving it largely intact. Unlike timber harvesting in Europe or North America, whole hillsides aren't shaved at once, and unlike deforestation for cattle ranches or palm plantations, the jungle is largely maintained. In fact, to the amateur eye, it looks almost unchanged. But it has been changed.

A commercial logger looks for high-quality wood for furniture or construction. They identify suitable trees before cutting and removal. But, on average, for every tonne of useful wood extracted, 6 tonnes are damaged or destroyed in the entire harvesting process. This ratio of waste to use can even rise to 20:1. Halving this collateral damage is the aim of 'reduced impact logging for climate' or RIL-C and, if implemented across the tropics, these practices could reduce our overall human-induced emissions by around 1.5 per cent.

So why does timber extraction leave such a wide trail of destruction? When the trees fall, they often squash others beneath. Some apparently desirable trees are discovered to be of poor quality after felling and not worth removing. Shifting the trunk across the forest floor smashes smaller trees in the way. Roads for the logging trucks cut long, wide ribbons through the woods. The Nature Conservancy is working in Mexico, Indonesia, Peru, Gabon and the Democratic Republic of Congo to do it differently.

Firstly, the logger should make a more accurate assessment of a tree's quality while it is still standing by doing a 'plungecut' where a chainsaw is driven into the base of the trunk to reveal if it is hollow without killing it. If it looks like viable timber then be sure your lumberjacks have the training and knowledge to drop the tree where it will cause the least damage to its neighbours. Most tropical trees are removed along so-called 'skidways' where vegetation is cleared to allow the trunk to be dragged to a track. Once again, these can be planned to be narrow and sensitive to other forest dwellers rather than a careless trail of destruction. Or use a 'logfisher', an adapted mobile crane with a very long cable that 'fishes' the timber out with less skidding. The logging roads themselves can be half the width – down from 30 to 15 metres – given better drainage, surfacing and construction techniques. Combine all these habits and many more trees are left standing to hold more carbon and potentially grow on to be valuable to the logger.

In Peter Ellis's words, none of these ideas are rocket science but when combined they can be surprisingly effective, reducing carbon emissions from logging by 50 per cent, and he is unapologetic about this practical, rather than idealistic, approach to conservation. He became inspired as a Peace Corps volunteer in Gabon in his twenties. 'I remember the moment when I stood out in that amazing, incredibly biodiverse, forest with the sounds of the birds and the animals all around me and I knew I wanted to pursue a career in forestry. These places need to be preserved and working with loggers we can do that. Loggers need to make a living and, if they don't, the alternative would be, and is becoming, much more destructive forms of land use like oil palm plantations – a desert of one species with much lower carbon retention.'

On the Indonesian island of Kalimantan, The Nature Conservancy's logging company partner Karya Lestari is proud of its concession that retains wildlife while generating an income. There are orangutans, civet cats, gibbons, hornbills and deer alongside occasional chainsaws. The organisation says RIL-C principles do take more time and cost in planning and training but the extraction operation itself is more efficient. This fine balance matters, says Peter Ellis. 'The first question you always get from an individual forest manager is "How much is this going to cost me? I'm worried about my bottom line." Logging is already a pretty low-margin industry and we are doing studies on cost/benefit. One thing for sure is that RIL-C saves lives. One in ten tropical loggers dies as a result of their own logging. The care and precision that goes into RIL-C makes it much safer.'

In Chapter 31, we establish that wood is better for climate change than concrete and steel, so sustainable felling should be environmentally benign in higher latitudes and near the equator. But tropical forestry has extra hazards, providing reasons to tiptoe not bulldoze. Firstly, the tropic's unparalleled wildlife and diversity deserve to be on a conservation pedestal. Secondly, trust: governance and oversight in many equatorial countries are poor. How can we be sure RIL-C principles are actually being followed and not just spoken about in conferences or written in deals to get carbon credits? Satellite imagery can help with the policing but Peter Ellis says The Nature Conservancy has started using on-the-ground auditors in Indonesia and are developing the National Gabonese Monitoring System in 2021.

I wield a chainsaw myself sometimes, felling trees and chopping logs for a wood-burning stove. I have, in effect, a concession in local woodland where the owner insists on a brand of reduced impact logging – no wheeled vehicles allowed, and only take the fallen or dead trees. She's protecting the bluebells, not climate, but the impact is the same: forest vegetation is preserved and fewer trees are felled as only those I can carry out myself, or load on the shoulders of family and friends, are taken.

Desirable destination

Cutting total greenhouse gas emissions by 1.5 per cent by the adoption of reduced impact logging principles.

How to get there

Policy: governments have to insist on RIL-C practices as a condition of granting or relicensing logging concessions.

Enforcement: trained auditors on the ground and satellite images from above checking compliance.

Certification: reliable guarantees that timber products are produced under RIL-C that customers can demand.

Fringe benefits

- More habitat for wildlife.
- Fewer injuries and deaths from logging in the timber trade.

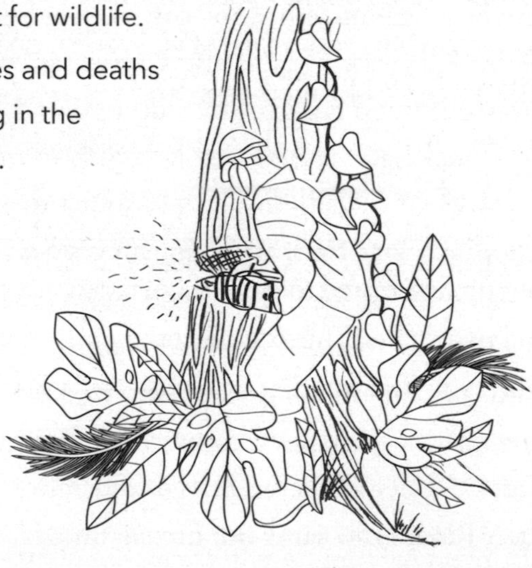

12
Bountiful Bamboo

As I look around the room where I am writing, virtually everything I see could be made of bamboo, and some already are. Desk and chair? Naturally. Computer mouse, keyboard and screen frame? Seen them all online. Books? Check. Doors and walls and beam and shelves? Easy, with engineered bamboo being stronger than steel. Clothes? Bamboo fibre makes soft, comfortable fabric all the way down to your underwear. I've even ridden a bamboo bike. All this from glorified grass and yet, further glory awaits as a noble ally in the fight against climate change.

Arief Rabik, almost literally, owes his life to bamboo. His mother needed emergency help with his birth and, while in labour, negotiated to pay the doctor with a giant bamboo sofa and two chairs. The story becomes only a little less outlandish when you find out that his mother, Linda Garland, did much to promote and glamourise bamboo for interior design, with celebrity clients such as David Bowie and Mick Jagger. Arief Rabik now runs the Environmental Bamboo Foundation in Indonesia, a charity dedicated to restoring land and capturing carbon with 1,000 'bamboo villages'. Each settlement will be surrounded by about 20 km^2 of bamboo forest mixed with crops and livestock. It's good for the soil, good for the local economy and good for the climate. He then wants to expand the idea to nine other countries. 'Collectively, they will absorb and remove out of the atmosphere 1 billion tonnes of carbon dioxide every single year.'

Bamboo is the fastest growing plant in the world. In tropical settings, with plenty of water, it can add 1 metre every day. In a week, it will be approaching full height and then puts on woody bulk and foliage before dying back after five to ten years. Such rapid growth ensures that bamboo absorbs CO_2 from the atmosphere much faster than trees, but its margin of victory depends on how it's managed and how the mature bamboo is used.

The most effective carbon sponges are bamboo forests planted with the high-yielding varieties in the best tropical conditions. Then, the fully grown stems are coppiced and made into durable products such as furniture or building material. This way the carbon is locked in the shelf, beam or bookcase. Bamboo canes, what you see above ground, are just a branch from the rhizome (like an underground tree trunk) and, when cut, new shoots quickly emerge to take their place and the cycle begins again. When managed in this way, a hectare of bamboo can take in around 50 tonnes of CO_2 every year as it grows.

If you grow and harvest the canes in the same way but assume that the bamboo products market is saturated or unavailable, then the next best thing is to turn the bamboo into biochar. This is made in a similar way to charcoal by heating up wood or bamboo to about 500°C (932°F) but without any oxygen so it can't burn. Some carbon dioxide is lost in the 'cooking' process but around 50 per cent remains locked into the biochar even when it is then spread on to farmland as a very effective soil improver, as explored further in Chapter 17.

A mature bamboo forest, without harvesting, reaches an equilibrium, five to ten years after planting, whereby CO_2 absorbed in growth is offset by that emitted from respiration and decomposition. But by then, when you combine stems, leaf litter and rhizomes below ground, the

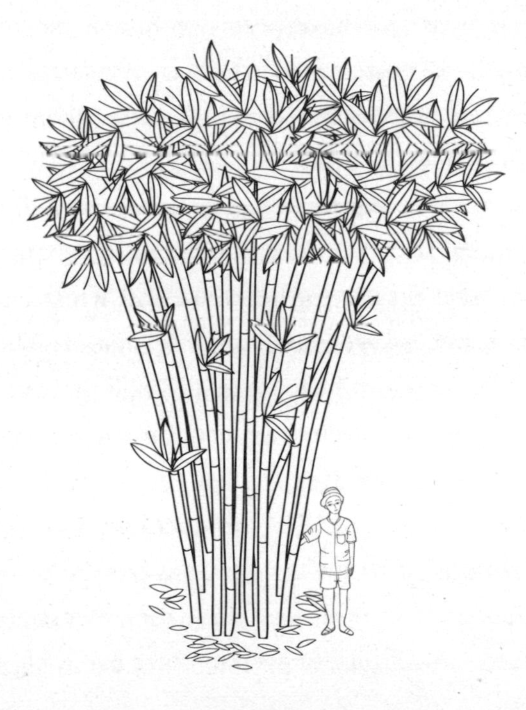

bamboo stand will be holding up to 400 tonnes of carbon per hectare. With at least 350 million hectares of land suitable for bamboo planting globally this would lock in up to 140 billion tonnes of CO_2; that's around three years' worth of total human emissions.

So how can we approach these stunning figures? Arief Rabik believes the answer lies in putting the climate change benefit towards the end of the sales pitch, and putting more immediate community benefits first: 'We call it S, E, E, in that order: social, economic, ecological. In our model with local management, you cannot have the bamboo without the village.'

It all rests on improving degraded land. These are often vast areas that have been deforested and then heavily cropped or grazed so that their vegetation and soils have become impoverished and at risk of erosion. The World Resources Institute says that, globally, 2000 million hectares, an area slightly bigger than Russia, is awaiting improvement. Bamboo is a land restoration champion. The rhizomes bind up the top metre of soil while also letting in more water, oxygen and organic matter. The canopy protects the soil from heavy rain and falling leaves become nutritious litter. Put these together and you can boost fertility while limiting flooding and soil loss. Nevertheless, Arief Rabik doesn't want a bamboo monoculture but mixed plantations with other native trees such as banyan or figs. In his village model, land is also set aside for other crops such as cocoa, coffee and fruits, or even to graze livestock. Contrary to perception in the West, bamboo is not necessarily invasive: most tropical species grow from clumps with six poles each and don't spread.

Regular coppicing of the bamboo also provides an income, with each village being able to harvest about 6 tonnes of poles every day. Around half of that is processed into strips for sale to construction or furniture companies and the remaining 3 tonnes is made into fuel pellets – a renewable energy that could power homes, small businesses or even biochar kilns. Arief Rabik then gets lyrical: 'This "Bamboo Village" vision is like the masterpiece of an oil painting. The bamboo is just the canvas because it can stabilise the soil and bring back the water. The artist is the ecosystem and then everybody else can benefit from this.'

He isn't a lone voice singing the praises of bamboo: in China it is a multibillion-dollar industry, India has a 'National Bamboo Mission' backed by the prime minister and many African countries are seeing its potential for both economy and ecology. But overall, it still seems to have an image problem: in the West it is often viewed as dangerously invasive in gardens, while in much of Asia it is still stigmatised as the as the 'poor man's timber', a relic of an impoverished past. Surely it is time to appreciate this miracle grass.

Desirable destination

Absorbing 2 per cent of our greenhouse gas emissions by 2050.

How to get there

Land restoration: widespread planting of bamboo on degraded land across an area the size of India.

Substitution: use bamboo in long-lived products such as buildings and furniture in place of cement and steel.

Partial combustion: bake in charcoal kilns and spread the resulting biochar on the land.

Fringe benefits

- Reduced soil erosion.
- Increased flood resistance.
- More income streams for rural areas.

13

For Peat's Sake

In May 1950, a body was discovered in the ground near the Danish town of Silkeborg. The male had a day's growth of stubble, wore a sheepskin hat, a leather belt and a noose around his neck. With well-preserved skin and intact facial features, police initially assumed they had found a recent murder victim. But they were out by over 2,000 years, fooled by the amazing preservative powers of peat – the same chemistry that makes peatlands critical to solving climate change. The 'Tollund Man' had barely rotted in the bog, and neither does the plant matter of millennia.

Peat is dead and partially decayed organic material that has accumulated over many thousands of years while soaked in water. In the tropics it's generally formed from leaves, roots and branches; in colder climates, the ingredients are more likely to be moss, herbs and shrubs. Peat bogs can be 1–18 metres thick and half their dry weight is carbon so, despite covering much less land than forests, they hold twice as much carbon. By draining peat, digging peat, burning peat and gardening with peat, we are letting that stored carbon escape. Around 4 per cent of our greenhouse gas emissions come from degraded peatlands. On average a hectare of dry peat in Ireland emits 6 tonnes of carbon dioxide per year – that's equivalent to the annual pollution of five cars.

It's the water that keeps the carbon locked up in peat by separating air from the organic matter and allowing the layers to build up over the centuries while largely sealed from decay. These soggy, somewhat spooky, landscapes held off human exploitation far longer than the forests, but the development of drainage technology changed all that. From the seventeenth century onwards, pumps and ditches started to open up the peatlands of Europe to farming. The black, crumbly soil was found to be the perfect growing medium, especially with added fertiliser, and the fens evolved from a treacherous marsh to a bountiful breadbasket, a trend repeated today in the tropics where formerly untouched bogs have been drained for palm oil plantations with the added threat of wildfires. In Indonesia in 2015, a massive blaze took hold in the dried peat and burned for months, at its height emitting carbon dioxide at the same rate as the entire US economy.

'The good news is that all you need to do to stop this is reattach the plumbing, put the water back in like turning on the tap and putting the plug in the bath.' So says Florence Renou-Wilson, a peatland expert from University College Dublin. She grew up in Brittany, France, but got sucked into Irish bogs the moment she was first shown one in 1995. 'Most people see wetlands as fetid and hazardous. But I was fascinated by their biology and mystery. I still am.'

Ireland's changing relationship with its soggy interior is a microcosm of what's happening worldwide. With few trees and scant coal reserves for fuel, many in rural Ireland turned to cut and dried peat to heat their homes. Some still do today. In the twentieth century, it became a valued natural resource with a state-owned company, Bord na Móna, industrialising its extraction and even using peat in power stations to generate electricity. It became a valuable export, too, much in demand from gardeners and growers in the UK. Added together, the degraded peat bogs of Ireland emit as much greenhouse as the country's transport sector.

But now, the availability of other fuels and environmental pressure have led to a complete reversal – in early 2021 the Bord na Móna not only stopped all further peat extraction but also promised to rewet the extensive areas it controls. Of the 1.4 million hectares of peat bog in Ireland, Florence Renou-Wilson reckons half will be rewetted in quite a short time. As for personal use, she thinks support from the government may be necessary to help the few who still rely on peat for home heating but she has little use for the 'it's part of our culture' argument. 'We used to say that about watching the guillotine in France. Sometimes you have to leave traditions aside.'

Stepping away from the Emerald Isle, most other peat-rich countries haven't yet had such a full switch from exploitation to protection but many are beginning to see peat rewetting as a potent nature-based solution to climate change. Block the drains and rainfall does much of the work for you, not restoring the bog to its virgin state but at least immediately arresting further carbon emissions. One third of tropical peatlands are in Indonesia and, after the fires of 2015, the government teamed up with environmental charities to set up the National Peatland Restoration Agency with the aim of rewetting 2.7 million hectares of degraded forest swamp. It's an expensive business largely reliant on donations and carbon offsetting credits as there isn't yet an international funding structure to pay for peatland rewetting. But there could be a way of squeezing some revenue from the restored bog.

Welcome to 'paludiculture', or farming on a marsh. Valuable plants can be grown and harvested on a waterlogged ground. In the northern hemisphere, reeds could be used as energy crops and building materials, black alder for timber. In the tropics, many economically useful species could grow in wetlands, including sago palms for a staple starch and the latex ingredient for chewing gum.

We need peat but spreading the message beyond committed academics, environmentalists and cutting-edge farmers is difficult. Unlike woodlands we still harbour such a negative image of wetlands: there is no boggy equivalent of 'tree-hugging' and politicians chant 'drain the swamp' as a euphemism for ending corruption. But there is one way you can show your love for peat according to Florence Renou-Wilson: go 'peat-free' at the garden centre. 'Every bag of peat you buy has come from a bog that is drained, polluting the water, releasing carbon and harming wildlife. It degrades an ecosystem critical to mankind. We need to hammer home the message that gardeners have no need to use peat whatsoever.'

Desirable destination

End the drying and degradation of peatlands by 2030, saving 4 per cent of our greenhouse gas emissions. Thereafter, allow rewetted peat to become a gradual carbon store.

How to get there

Policy: follow Ireland's example and make peat degradation illegal.

Natural: allow some fens to return to their natural state.

Agricultural: perfect paludiculture farming methods that yield some economic return from the peatlands.

Financial: support verifiable and robust carbon-offset schemes that pay for the rewetting of tropical peatlands.

Fringe benefits

- Better habitat for plants and animals.
- Improved natural filtration for drinking water – water companies will often pay for peat restoration.

14

Ocean Farming

The closest I have ever come to flying – outside of an aircraft – is swimming off rocky Hebridean coasts. Face-masked and peering down through crystal waters, my favourite is gliding over towering kelp beds: swaying seaweed forests harbouring crabs and scallops, occasional seals and imagined sharks. I feel out of my world with a tingling mix of pleasure and unease. But this world could play a big part in saving ours.

About half of all photosynthesis, the foundation equation for life, occurs at sea. Seaweed and micro-algae take up carbon dioxide and release oxygen just like plants on land. And, just as tree planting is seen as a climate change solution, so is seeding seaweed.

It's estimated that seaweed already locks up around 634 million tonnes of CO_2 every year in the ocean, which is a little over 1 per cent of our annual human-made emissions, and there is huge scope for growth. But it's *using* seaweed at scale that delivers the biggest climate wins: it can replace so many carbon-intensive processes in farming, fuel production and plastic manufacture. It's estimated that by just using a thousandth of the ocean area to grow seaweed we could provide one quarter of human protein demand and use 6 per cent less farmland. Seaweed is currently farmed on a large scale in parts of Asia and it's commonly seen in oriental cuisine but it's also used in everything from toothpaste to ice cream, face cream to fire retardant fabric. Demand for seaweed as a meat alternative is growing strongly; it is a source of non-animal protein that avoids the deforestation issues around soya and delivers the savoury umami flavour carnivores crave. But a thousandth of the ocean is 0.5 million km^2, a fifteenfold increase on the current farmed area.

Bren Smith, the founder of marine aquaculture company GreenWave, believes we can do it. Based in New England, USA, his farms grow kelp alongside, mussels, scallops and clams all in the same place. He calls this polyculture and says he's recreating what nature does on land or sea: growing different things at different levels. A three-dimensional web of anchors, ropes and buoys mimics a natural reef ecosystem where he can produce things that 'don't swim away and I don't have to feed'. For each hectare block – 100 × 100 metres – he can grow enough kelp to make 9 tonnes when dry and yield around half a million shellfish. The kelp can become animal feed or fertiliser and the seafood finds its way to shops and restaurants, while the growing seaweed itself is helping to take CO_2 and excess nitrogen from the water. But there is yet another benefit: the kelp beds provide a nursery and a haven for other marine life, helping to increase the wild abundance of the ocean.

GreenWave has trained 120 ocean farmers and has thousands more on the waiting list. But Bren Smith admits it's tough: 'You can't easily see what you grow and the water is always changing. In the first few years I managed to kill millions of oysters. But eventually I developed a "blue thumb"' – rather than green fingers.

He envisages hundreds of marine polyculturists dotted along the coast, each with their own small plot of seabed and water column above, like fishing communities of old but now working with the sea, not plundering it.

Five thousand kilometres across the Atlantic, at the mouth of Loch Linnhe on the west coast of Scotland, you find another seaweed farm. This one is run by the Scottish Association for Marine Science (SAMS), and their tiller of the deep is Dr Adrian Macleod. He's also passionate about kelp, even slipping it into his miso soup at home, but with an academic's demand for evidence in place of entrepreneurial zeal.

As we skitter across the wave tops in his no-frills plastic speedboat he explains the challenge: 'I wouldn't be here talking to you if I didn't see a massive future for large-scale seaweed farming. But it's incredibly important to be honest when it comes to climate change – are we really sequestering carbon?'

We arrive at the farm site and Adrian hauls up a rope laden with kelp and so much else: this is no monoculture as the seaweed provides an anchor and a haven for an abundance of sealife.

As seaweed grows it absorbs dissolved carbon dioxide from the water. This is held in the flesh of the plant and in slimy sugars, which often coat the brown gelatinous surface. Throughout the kelp's life these sugar molecules and bits of the seaweed itself are washed away or broken off by storms. Much of this material tumbles to the seabed, away from the atmosphere, and some falls to the deep ocean floor where it is permanently sequestered. Recent research suggests that around 10 per cent of the carbon taken during growth ends up in long-term storage.

But Adrian Macleod points out that currently seaweed farming involves extra carbon emissions from boat fuel, plastic buoys, concrete anchor weights and steel hawsers. His team at SAMS are working to tip the balance favourably by using more climate-friendly raw materials and aquaculture techniques such as spray seeding the ropes with microscopic juvenile kelp. He can see collateral benefits right now flowing from seaweed farming for the marine ecosystem and the economic health of remote coastal communities; large-scale carbon savings are visible but a little further off. A neat example of imaginative thinking to create efficiencies is siting seaweed farms between windfarms where they already have the service infrastructure and trawling is off limits.

There is no doubt that a massive ramp up of seaweed farming faces challenges: where are the regulations, where is the finance and where is the market for all that seaweed? But big global organisations see the potential and are seeking solutions: the United Nations, the World Bank, Lloyd's Register Foundation.

Bren Smith believes the sunlight falling on our inshore waters is an asset we can no longer afford to ignore. Both for our world and the blue world, protection is no longer enough: 'You could turn the whole ocean into a marine park and it will still die in an era of climate change. Conservation, on its own, does not deliver. We need new ways to regenerate our oceans. Seaweed farming is a big part of that.'

Desirable destination

Absorbing 2 per cent of the world's greenhouse gas emissions from seaweed farms in 0.33 per cent of the ocean. This would be a 600-fold increase in current seaweed farming.

How to get there

Technology: developing cheaper and lower carbon aquaculture gear and underwater unmanned 'farm vehicles'.

Policy: massive incentivisation from government and improved regulation and management of inshore waters to allow seaweed farms to develop across wide areas. Seaweed farms are highly compatible with windfarms and could share the same seabed. Need to resolve likely opposition from the traditional fishing industry.

Skills: a huge training programme in ocean farming.

Markets: developing the appetite for seaweed in human food and animal feed, and as an industrial raw material.

Fringe benefits

Better food: seaweed and the seafood grown alongside would provide a vast healthy and low-carbon food source.

Better habitat: the seaweed farms would become a refuge and a nursery for wild marine species including fish and shellfish.

Better jobs: provision of high-skilled jobs in aquaculture, processing and associated industries in remote coastal areas where employment is scarce.

Farming

15

Zero-carbon Crops

In early February 2021, I am looking over a rolling 20-hectare field in Northamptonshire. Snow covers the treacly mud but not the sprigs of stubble remaining from last year's barley and the hardy leaves of this year's emerging oilseed rape. Laid out like a flecked tweed blanket, it looks attractive but not extraordinary: a pleasant winter scene on a typical arable farm. But this is the birthplace of a revolution: living and growing proof that farming can swap sides in the fight against global warming to be one of the climate's greatest allies. And it can do so while still producing plenty of food.

The farmer is Duncan Farrington: 'This field is absorbing enough carbon dioxide out the atmosphere every year to offset the emissions from around 200 medium-sized family cars on UK roads every year. Across the whole 270-hectare farm it adds up to about 2,700 cars or 500 economy flights round the world.'

Farming is a huge part of our climate change problem so switching it to the solutions side would be a very big deal. Land use, including forests felled for agriculture, accounts for around one quarter of our greenhouse gas emissions. The actual practice of farming contributes about half of that, and despite the publicity around cows, growing *crops* is as threatening to the climate. It is about the soil and chemical fertiliser. Disturbing the ground through planting or ploughing allows soil carbon to oxidise and escape as carbon dioxide. Making fertiliser releases a lot of carbon dioxide and, once spread on the field, the main ingredient, nitrogen, combines with oxygen to form nitrous oxide: a greenhouse gas with a potency 300 times that of CO_2.

Priority one is to reverse the loss of soil carbon. Duncan Farrington and a growing band of so-called 'regenerative farmers' believe it's all about having plants do more work and machines do less.

In any crop, only a small proportion of the plant is food for us, such as the ear of wheat or the pea in the pod, so it's best to take the foodstuff away but leave the leaves, stalks and roots to rot down in the field. Then quickly sow something else as bare soil is a waste of the sun's photosynthetic power to suck up CO_2 in a leaf. After harvest, this could be a cover crop mix such as oats and vetch or, after planting, a companion such as buckwheat and clover – something that complements your cash crop but doesn't compete with it. That too dies back and adds further compost. All these plants have roots, which also perish and deposit precious carbon in the subterranean bank.

The next trick is to achieve all this with as little tractor traffic as possible: so-called min-till (minimum tillage).

Don't use a plough to break up the soil as the worms should be doing this for you if they have enough organic matter to chew on. Where possible, reduce wheeled visits in the field to avoid the soil becoming compacted by heavy machinery. This is good for the ground, the farm budget and the carbon budget: Duncan Farrington says his fuel use has dropped by 60 per cent.

He then digs a clod out of his field and, even after a sodden winter, it has an open crumbly texture with old roots, worm holes and worms. 'We travel lightly on the field and work with nature. The worms and the bacteria are providing nutrition for the plants and there's roots that have been there for the last 10 to 20 years and that's all carbon.'

Duncan then extracts a graph from his pocket and reveals, between the muddy smears, that his soil carbon has increased by more than 75 per cent in the last 18 years, absorbing 5,000 to 6,000 tonnes of CO_2. There is no sign yet of the line flattening out. But the graph also has another line, one measuring soil fertility, and this rises in sync with the carbon. The added fertility from the organic matter also means he is using less artificial fertiliser and cutting down the associated greenhouse gases as well as his chemicals bill. So, far from having to compromise on profit, he can have a healthy balance in the bank and bulging carbon store in the field: sustainable for the business and the world. This is what makes his story so compelling. Duncan runs a profitable arable farm producing the high-end culinary rapeseed oil Mellow Yellow.

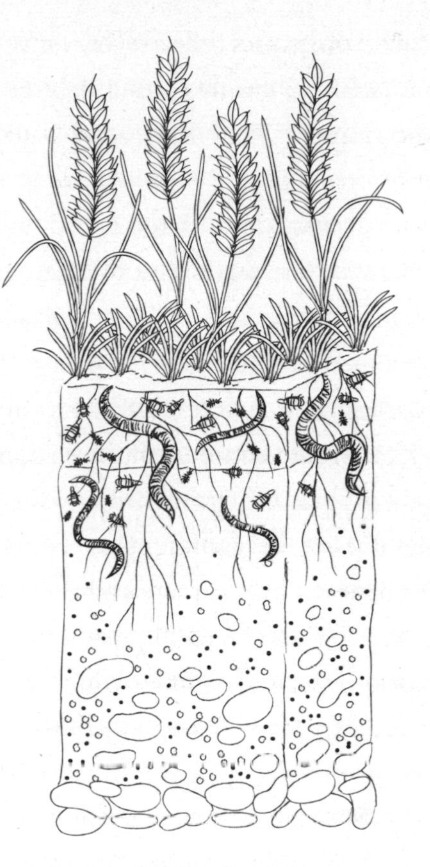

He is not, in his own words an 'eco-nut', but believes fervently that this approach could work well across commercial agriculture in the bread baskets of Europe and North America. 'It is suggested that if global agriculture was to farm in this manner, there is the potential to remove between 10 per cent and 30 per cent of global carbon emissions annually. And that is the equivalent of removing all global emissions from transport – planes, ships, trains, cars. I am showing that we farmers can very much be part of the solution.'

This approach, sometimes called regenerative agriculture or conservation farming, clearly owes a debt to the knowledge of organic farmers. They have appreciated the value of vibrant soil for years and it is no coincidence that their biggest membership group in the UK is called the *Soil* Association. But regenerative farming is a more pragmatic method, lacking the dogmatic rejection of chemical additives and focusing on the long-term survival of farming and the planet.

Duncan Farrington is proud of what he's achieved and has had some high-profile admirers in his field, not least the Chinese minister of agriculture and their envoy to the United Nations. But he doesn't feel farming gets the recognition it deserves in the drive to halt climate change. 'Something I have found frustrating over 15 to 20 years is that people like a big shiny solution, something to look at that's taken a lot of technology and expertise. Whereas here there is nothing to say, "Look we've invested all this money and invented all this technology." You can't see it, but the worms and the roots, they're doing it. We're standing here on a cold sunny bright day and it's happening beneath our very feet and will keep happening. Nature is doing it every day.'

Desirable destination

By 2040 arable farming will be carbon neutral, cutting 10 per cent of greenhouse gas emissions. In the following 20 years it will sequester hundreds of gigatons of carbon before reaching an equilibrium with maximum soil organic matter by the second half of the twenty-first century.

How to get there

Robust data: reliable metrics of soil carbon enabling so improvement can be measured.

Policy: farming is hugely supported by government funding across much of the world so this funding can be made conditional on a carbon-cutting trajectory. End tax cuts on farm diesel.

Knowledge transfer: dissemination of regenerative farming practices across the industry.

Retail: reliable information on the carbon footprint of food so shops and customers can choose to buy the climate-friendly product.

Fringe benefits

- Less money spent on fertiliser.
- More farm wildlife.
- More nutritious food.

16
Dry Rice

When does the life of a plant begin? You could argue that a seed dwells in a state of near eternal suspended animation, but surely the vital spark is evident as the kernel cracks and the plant grows: extending up towards the light and down towards the drink. This is the moment that animates Dr Smita Kurup. Germination is her field and her key to stemming a huge source methane: a potent greenhouse gas.

Rice growing causes around 2 per cent of global warming, approaching that attributed to flying. On average, a large bowl of rice has a similar climate impact as driving a car for a couple of kilometres. Most of the world's rice comes from paddy fields, small plots flooded with about 10 cm of water. The role of the water is to supress weeds; it is just a big wet blanket that smothers everything bar the rice. But, sadly in terms of climate change, it also keeps oxygen in the air away from the soil. This means the breakdown of plant material happens without oxygen: it is anaerobic and that creates methane, also known as 'marsh gas'. As paddy rice grows across 161 million hectares, we have created a huge bog about twice the size of France with methane escaping to thicken our global warming blanket. A molecule of methane traps heat 25 times more effectively than carbon dioxide. It is the world's super-insulator and we have more than doubled the amount in our atmosphere compared to pre-industrial levels. The other big methane sources are cows (see Chapter 17) and leaks from the gas industry. To staunch the flow of methane from rice cultivation you need to consign the paddy field to history and that is Dr Kurup's dream.

She works with a team at the plant science hothouse of Rothamsted Research, just outside London, trying to create a variety of rice that thrives in a conventional field: so-called dry seeded rice (DSR). The idea of growing rice in this way is not entirely new: the International Rice Research Institute based in the Philippines has been doing field trials for some years. But current rice varieties, bred for success in the paddy field, do not do well in a DSR system.

Dr Kurup's passion is for giving plants the best start in life. She grew up in Delhi fascinated by the living world and focused on plant science because she was 'a little squeamish about cutting up animals'. After moving to the UK, she studied germination in poppies, cress, rapeseed and now rice, giving her a unique insight into the 'spark of life' moment when a plant's nutrition comes not just from the seed but from water and light. Her team and partners have a simple aim: to breed a high-yielding, hardy rice variety that grows well in a dry field system. And the importance of the work is not lost on her. 'The possibility of changing something swiftly in the real world is a massive motivating factor but it also gives me a sense of responsibility, not to mess this up. But my confidence level that this will succeed is around 95 per cent. I do believe the world of rice will be revolutionised by Rothamsted and our partners.'

The revolution is being planned in a series of greenhouses and labs. Dr Kurup takes me to a warehouse humming with rows of what look like big steel fridges but these are holding warmth, not cold. She opens one up to reveal rice seedlings enjoying the heat, light and humidity of Southeast Asia. She admires them with all the pride of a parent: 'Our germplasm is looking beautiful; all my babies are thriving.'

She then compounds the delighted mother role by showing detailed pictures of rice seeds' earliest development: slightly ghostly images of stringy roots and striving shoots against a black background, the plant scientist's version of the in-utero baby scan. But this is about so much more than pride; these high-resolution images are key to the success of the project.

As a seed germinates, it reaches down with roots for water and up for light, starting with something called a mesocotyl: the first fleshy upward protuberance, which then becomes a shoot. A paddy-sown seed leads a pampered early life: surrounded by water, it has no need for rapid root growth and, being on or near the surface, light is close by. If you plant such a seed in a dry system, lacklustre root devel-

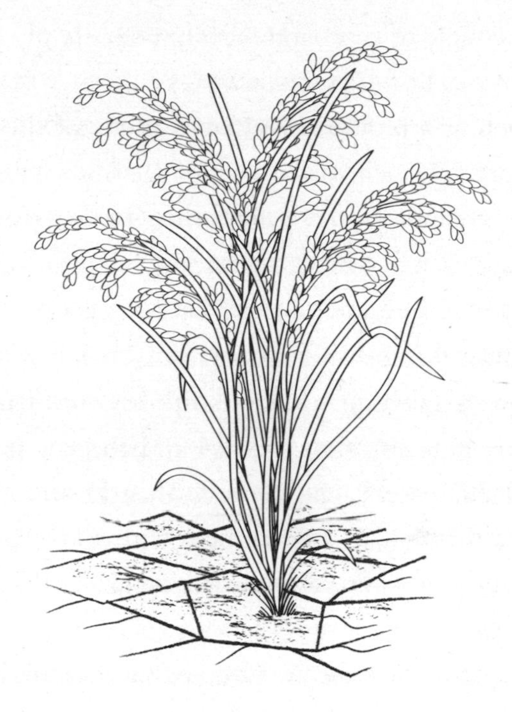

opment means it does not get a drink and its slow upward growth doesn't reach the light quickly enough. A dismal fate awaits. But Dr Kurup discovered little-known varieties in the gene bank of the International Rice Research Institute that *do* grow robust roots and a long mesocotyl. The first stage of the research involved planting around half a million seedlings from 650 varieties. From those she has selected the ten most promising to develop and cross with existing high-yielding breeds. At each stage, her partners at the International Rice Research Institute are field testing her varieties in the Indian state of Punjab and near Varanasi in Uttar Pradesh. Dr Kurup believes they will have farm-ready seeds in a couple of years and within a decade most of the world's rice will be grown this way.

But there is a problem. Unlike so many of the world's commodity crops such as wheat, maize or soy, a lot of rice is grown on smallholdings not big farms. The average paddy farm in Asia is less than 2 hectares. Changing the farming methods of hundreds of millions of people is harder than a smaller number of big operators. Dr Kurup says the paddy growers' farming may be traditional but their digital connectivity is twenty-first century and the key is to find their local influencers – people who they know and trust. Partners at the Punjab Agricultural University have even made a video featuring a farmer singing about the benefits of dry seeded rice.

The big attraction for the farmers has more to do with increasing profits than reducing global warming. Paddy-grown rice is becoming unsustainable in many places. It needs lots of labour and lots of water, and across Asia these resources are becoming scarcer. It also harms the soil structure. If DSR can deliver increased yield at lower cost, it sells itself, with limiting climate change thrown in for free.

Desirable destination

Reduce emissions by 0.5–1 per cent by widespread adoption of dry rice techniques.

How to get there

- Continued development of highly productive strains adapted to non-paddy systems.
- Promotion of dry rice-farming techniques by local influencers and governments.
- Climate-impact labelling of rice, allowing consumers to favour dry rice.

Fringe benefits

- Less demand for scarce water.
- Less demand for scarce labour.
- Improved soil quality.

17
Degassing Cows

Dead or alive, I like cows. Their meat tastes great, their milk is nutritious and the consequent cheese is a wonder of the edible world. They even look good: always ready for their close up with big brown eyes and lengthy lashes. Much of the world has fallen for them: the global herd numbers 1.5 billion cattle, that's one for every five people. But this is a toxic relationship as, in such numbers, they are harming the climate that sustains us. So, must we end the affair or can our sweetheart change?

When it comes to climate change, cattle have three main flaws. The way they digest grass in the rumen and three further stomach compartments results in methane emissions from front and back ends. Each molecule of methane has a global warming power 25 times that of carbon dioxide, although it doesn't remain in the atmosphere so long. More methane escapes from the dung along with nitrous oxide, another potent greenhouse gas. Cows are also markedly inefficient at turning their feed into body mass – the meat we eat. This matters because it means they demand a lot of land to graze or to grow their supplementary food, land that could otherwise do a more climate-friendly job such as grow trees and, in some cases, native forest is replaced by pasture. It's calculated that rearing cattle and buffalo combined with growing their feed is the source of about 10 per cent of greenhouse gases, a figure that rises if you factor in deforestation.

Despite, or perhaps, because of that hefty charge sheet, Professor Eileen Wall, climate expert with SRUC, Scotland's Rural College, says scientists need to work with livestock farmers, not shun them: 'Red meat is going to be part of the global food diet for 50, 100 years and more so we have to do something to help farmers get new technology and farming practices up and running. We [society] need to help farmers be less demonised. There are so many activities happening that I do believe we'll get close to net zero [beef] in the next 20 to 30 years.'

Those activities fall broadly into five groups: grazing, feeding, breeding, digesting and housing.

Grazing: if your fields are a carbon sink it offsets some of the global warming effect of your herd above. Soil holds twice as much carbon as the atmosphere and around one fifth of that is held in pastureland. As it is below ground, its storage is more permanent than plant biomass. In a grassy field, most of the carbon is held not in the visible leaves but in the roots and organic matter of microorganisms and fungi. Husbandry that helps those roots, bugs and mycelium to grow will increase soil carbon. More farmers are practising 'mob grazing' where grass is left to grow longer before being munched intensively for a short time, then the animals are moved on. It is thought to mimic natural foraging where herbivores are shifted around by predators: the wolf pack is replaced by the electric fence. Taller grass has deeper roots, and more stems tend to get squished into the soil by so many hooves. These types of grazing, coupled with using mostly organic fertiliser such as manure and plant matter, can boost soil carbon rapidly. But it can't go on forever as there is a working limit for soil carbon: after around 30 years of rapid improvement your store would probably be full.

Feeding: varied diets react differently with the cows' gut microbes and produce different levels of methane. It is broadly true (and disappointing for the supporters of outdoor-reared beef and dairy cattle) that grain-based feeds such as maize or soya produce less methane per kilo of beef than grass, although that balance is tipped somewhat if you account for greenhouse gas emissions where the grain was grown or practise the mob-grazing system mentioned above. Researchers in Australia, New Zealand, the UK and the USA are all working on breeding new varieties of grass that would produce less methane when digested. It is still in labs and trial plots but if successful would really be 'green, green grass'. Feed additives such as certain seaweeds, oils and tannins have shown promise and one Dutch company claims that a quarter teaspoon of its product can reduce methane production in a cows' rumen by 30 per cent.

Breeding: when it comes to methane production, not all cows are created equal. In a typical herd the variation between the gassiest and the greenest could be 20–30 per cent. Gut microbes are strongly associated with a cow's genetics and therefore heritability, so it should be possible to breed out the worst climate offenders. I'd like to see cattle shows awarding rosettes for the lowest methane beast, not just the one with the finest rump. Breeding for rapid growth is a climate plus for beef cattle as the quicker they reach slaughter weight, the less time they have on the earth to belch methane into our atmosphere. Eileen Wall says this advance alone has already reduced methane emissions from UK cattle and can quickly go further: 'Our studies suggest that selecting animals that are more efficient could reduce their environmental footprint by 24 per cent in the next ten years.'

Digesting: if you want to get picky, it's not the cow itself that produces the methane but the microbes in its gut and further teams of scientists are targeting them. One approach uses antibiotics to inhibit the methane-producing bugs called archea, another is to add 'good' bacteria in the form of a probiotic supplement. Teams are also working on genetically modifying the bugs themselves to reduce their methane emissions, which is energy lost to the cow, and divert that energy into swifter growth. When and if these GM ideas move from the lab to the field, opponents of the technology will have to judge if the perceived risks outweigh the climate rewards.

Housing: could a cow's 'bad breath' be captured before it escapes into the wider atmosphere? The Zero Emissions Livestock Project (ZELP) is working on a wearable device that clips to a cow's nose and captures some of the methane as it emerges. It's early days but ZELP claims that its device traps about one third of the methane and it is working to improve that. The methane is oxidised to CO_2 so some advantage will be lost but it could still help to cut the climate impact even if it does not eliminate it. Many cattle, especially dairy herds, spend much or all of their time indoors. Eileen Wall's colleagues at the SRUC are looking at fitting ventilation units that can scrub the methane from the air as it leaves the shed and also looking at combining manure-handling systems with anaerobic digesters to provide biogas fuel for the farm: 'Micro digesters are now being made compatible with farms sheds of the future so they could convert manure into energy for use on the farm or selling back to the grid.'

The animal husbandry changes to achieve more climate-friendly cows could have controversial side effects: less time outdoors, swifter growth, gene editing and earlier death. It is an uncomfortable truth that skipping around on the hillside may be nice for the cow but questionable for the climate. Faced with all these dilemmas many people have decided the simplest way of cutting the climate impact of cows is to frequently or totally avoid beef and dairy. Eileen Wall thinks that is a respectable individual choice but in the meantime we need to change farming: 'There are so many solutions ready to go now, let's help farmers get those up and running. Cows and sheep are a key part of our agricultural landscape. It is fundamental that we use these lands that cannot produce human edible protein and energy to produce diets that will help feed the world healthily and sustainably.'

Desirable destination

Around 10 per cent of total greenhouse gas comes from cows; by 2040 that could be cut by half by reducing the emissions from managed herds in Europe and the US.

How to get there

Policy: in most developed countries farming is supported by government subsidy so for cattle farmers these should become conditional on carbon-cutting husbandry.

Retail: supermarkets and food producers should label the climate footprint of their meat so consumers can choose the low-impact option.

Research and knowledge transfer: improving effectiveness of methane-cutting innovations and spreading best practice throughout livestock farming.

Consumption: eating less meat.

Fringe benefits

Less air pollution and odour from more controlled use of manure.

18

Dash to Ash

Josiah Hunt's language is earthy – both in the literal sense, as the climate change solution he's delivering is all about the soil, and, also, he's partial to the Anglo-Saxon for emphasis. 'I was reading an article in *National Geographic* magazine back in 2008 about the state of our soils and realised we are so f****d. Population is rising, soil is vanishing, the planet is warming and we are accelerating how we are screwing ourselves.'

He'd studied agro-ecology and the soil carbon cycle. 'I was looking for a way that we could unscrew ourselves. How can I save the planet and get paid for it? Luckily, the same article mentioned biochar and I thought, "Holy Cow! This is the missing piece." How come this wasn't part of our soil science? It just smacked me round the face.'

Biochar is a type of charcoal optimised for agriculture. It is made by baking plant material such as wood, straw, leaves or food waste with very little oxygen. The restricted air means the material can't burn. About half the CO_2 and some flammable gases are given off and you are left with fragile chunks, which are 70 per cent carbon alongside some nitrogen and oxygen. Josiah Hunt made some in his backyard in Hawaii by bartering wood for work at a local sawmill. Then he spread it on some trial plots on a small holding. 'The results were significant – a lot of corn versus no corn.'

A revelation to him but a no-brainer to past generations of indigenous farmers in the Amazon basin and, quite probably, many of our agricultural ancestors. Tropical forest soils, once cleared, are not very fertile. The high temperature and humidity mean organic matter rots rapidly and the goodness is lost to the atmosphere. But parts of the Amazon are marked by an amazingly productive soil called *terra preta* – black earth. Analysis reveals remarkably high levels of charcoal, which in places dates back around 2,500 years: the native farmers were using it as a soil improver. In temperate-zone farming, the importance of charcoal was less appreciated as the soils tend to be more productive but it is highly likely that cinders from the fire were frequently mixed with animal and human waste before spreading on the soil. So how does it work?

It provides the perfect habitat for some of the things we value in a good soil: microbes, nutrients and water. Josiah Hunt credits biochar's performance to its extraordinary internal surface area: 'It's like a plant frozen in time. All those pores and tubes in the plant, the xylem and phloem where water once travelled, are home for masses of microorganisms. A small handful of biochar has a total surface area bigger than a basketball court.'

Like many more traditional farming practices it became marginalised by the dash to use synthetic fertilisers developed in the 'green revolution' of the twentieth century. But now it's returning with a vengeance because of what it does for the climate: biochar stores carbon. Half the carbon absorbed by the plant as it grew is locked up in the biochar for thousands of years. You want carbon capture and storage without giant industries? This is it.

Most agricultural land could accept some biochar, and it's been estimated by a group of researchers called Herculean Climate Solutions that if we spread a layer about 1.5-mm thick on all farmland, that's about 5 tonnes per hectare, we could absorb 29 gigatons of CO_2 every year: more than half our annual human emissions. That would require a biochar industry on an epic scale and growing dedicated feedstocks such as bamboo or eucalyptus, but even just using the waste products from farming and forestry to make biochar could absorb 2 per cent of our greenhouse gas.

Josiah Hunt is stoked. 'We have a way now to take carbon that was taken in by plants, held in their bodies. We can stabilise that in charcoal and bury it in the ground in ways that improve water conservation, nutrient management and crop yield, helping solve climate change and providing us with food security. It answers really big problems.'

His business, Pacific Biochar, is based in California where vineyards are a valued customer, and biochar can help with another problem too: wildfires. One of the reasons they have been so severe lately is that protected forests have become choked with too much young growth – in the past this would have been cleared by occasional, smaller, natural fires or deliberately burned off with indigenous techniques (see Chapter 25). Thinnings from fire prevention would provide the perfect biochar feedstock.

Given all these advantages and the fact that, in regenerative farming circles at least, biochar has been touted by some influential figures, why hasn't it caught fire? A question that raises Josiah Hunt's temperature. 'For the last ten years we've had people telling us the biochar industry isn't doing very well, it's kinda pathetic. Well, you're asking us to f***ing solve climate change but you are only paying us to grow peas. They just haven't been willing to pay for it … until now.'

What has changed is eligibility for 'carbon dioxide removal credits'. The production of biochar has been recognised by a number of certifiers in Europe and the USA as a robust carbon-removal technology. This means companies that are seeking to offset their emissions, which could come from anything like heating their building to transporting their goods, can pay for biochar that stores equivalent carbon. The state of California has its own carbon-offset programme worth billions of dollars. It's given a boost to fledgling biochar companies on both sides of the Atlantic. Some environmentalists question the whole offsetting concept as a possible excuse for inaction by lazy polluters but, as a means to kickstart what could be a massive carbon-sucking industry, it seems necessary right now.

For Josiah Hunt, it's transformed Pacific Biochar from a 'bootstrap company': 'Companies are now lining up to buy our carbon removal credits. We have just received an investment of millions of dollars and will be able to grow our output fivefold. In the past the price the farmer paid for the biochar had to cover the whole cost of production and many doubted whether it would give them enough return on investment. Now we can slash our prices it's an easy sell. There will be further massive scale-up unleashed by us having a financial reward for climate change mitigation, proving this whole biochar thing does have legs.'

Desirable destination

Make biochar from forestry thinnings and offcuts alongside crop residues and wood harvested from dedicated plantations of bamboo or trees to absorb 4 per cent of human emissions.

How to get there

- Use the carbon-offset market to kickstart the growth in biochar production and use.
- Divert as much waste biomass as possible into the biochar production chain.
- Educate farmers and land managers on the benefits of using biochar.
- Develop biochar on an industrial scale with plantations created across millions of hectares.
- Have biochar officially accredited as a BECCS (bioenergy with carbon capture and storage) solution.

Fringe benefits

- Better crop yield and tree growth where biochar is used as a soil improver.
- Improved nutrition from crops.
- Better soil quality, improving both flood and drought tolerance.
- Reduced risk of forest fires.

19

Phenomenal Photosynthesis

One equation underpins existence, but it could be improved. Photosynthesis, it turns out, is actually rather flawed: plants convert around only 1 per cent of the sunlight they meet into chemical energy. A growing crop of plant scientists think they could do better. Given that we are talking here about the miracle that turns light into life – a miracle unrepeated throughout the known Universe – this might sound like ingratitude mixed with hubris. But, if successful, it could help halt climate change with a second green revolution.

Photosynthesis is the transformation of water and carbon dioxide, in the presence of light, into a carbohydrate, which builds the plant, and oxygen, which is released into the atmosphere. For those that love an equation: $6\ CO_2 + 6\ H_{2}0 + LIGHT \rightarrow C_6H_{12}O_6 + 6\ O_2$. It happens on land and sea, and without it there would be no plants, animals or us. In fact, virtually no life at all.

In a greenhouse in North Yorkshire, photosynthesis is getting an upgrade. This is the field trial site where tomatoes are having 'sugar dots' added to their water supply, which could increase their yield by up to 20 per cent. Glaia is the company behind this project, which was initially developed by the University of Bristol, to enhance photosynthesis. Two of the brains behind it, David Benito-Alifonso and Imke Sittel, are admiring and tasting the ripe fruit. Then they take turns to explain how it works.

When sunlight hits a leaf, it excites molecules of chlorophyll which, in Imke's words, 'bounce around' until they meet a complex of other molecules known as a reaction centre where their energy can be converted into chemical energy – the carbohydrate. That is photosynthesis. Too many bouncing chlorophyll molecules can damage the plant so it uses other chemicals to calm them down safely but without contributing to plant growth. The process is known as 'photo-chemical quenching'. But these energy-absorbing chemicals stay around longer than necessary and 'waste' potentially productive light. It's a bit like going downhill on a bike, squeezing the brakes when it gets a bit hairy, but then finding they stay on too long, even when you are back at a good safe speed, and that gravitational help is wasted. Glaia's 'sugar dot' enables the plant to release the brakes on photosynthesis faster or, as it's known in the academic literature, to accelerate the recovery from photoprotection.

What is the sugar dot? It is extremely small – a nanoparticle, meaning it's a few millionths of a millimetre across – and Glaia says its precise chemistry is a trade secret. But the company insists it is safe and found naturally in honey, coffee, caramel and toast. Manufacture of the sugar dot is cheap and requires little energy; it is water soluble and can be applied through the irrigation system or with a leaf spray. It isn't a fertiliser or plant food as it contains no nutrients, and it doesn't demand more fertiliser to be effective. In all the trials, treated and non-treated plants were grown in identical soil so it appears the enhanced photosynthesis process makes better use of *available* nutrition and light.

The Glaia team say the sugar dot could be applied to a whole range of crops as it is water soluble but that doesn't necessarily mean it will deliver a bumper harvest. The extra photosynthesis could just mean bigger leaves or a fatter stem, and if you are growing fruit or cereals, that doesn't help food production. Laboratory tests with wheat, though, have been promising as photosynthesis appears improved and the extra chemical energy is stored in the grain. In fact, they saw yield gains of around 20 per cent, which is an impressive leap. But their first trials in commercial growing conditions are with soft fruit under glass. Here, inputs from soil, water and even light can be measured and, to some extent, controlled giving a clearer picture of the outcome. It is too early to have firm data from these trials but David and Imke are pleased by what they can see as the treated tomato plants are taller, and I can vouch for no loss in flavour as, with their blessing, I ate some results.

Growing more food is an important end in itself as the world population is expected to peak at around 9.5 billion by 2050 – adding another 2 billion souls to the current total. The massive growth in farm yields we saw in the latter half of the twentieth century with the 'green revolution' has slowed and much of that was achieved with heavy use of chemical fertilisers. Getting more food from the same land or less *without* environmental downsides is the now the goal. It is known as 'sustainable intensification' and enhanced photosynthesis could be critical to its success. Sparing land from food production means it can be used to absorb carbon in woodlands, wetlands and well-managed grassland.

Imke Sittel thinks shrinking our food footprint would be a huge step to limiting climate change: 'My biggest dream is actually contributing significantly to stopping climate change by going into a global market for the staple crops.'

And she believes her sugar dots could help with natural carbon storage directly, not just super plants but super trees: 'We could also potentially help reforestation. If we need to reforest an area and we can make the trees grow faster and reforest bigger, it's a whole different approach to help fight climate change. We just need a second green revolution.'

These are hopes not promises. The Glaia team are not claiming some panacea solution but just one among many that could help us cut climate change. Nevertheless, it is an idea that tinkers with the fundamental equation for life on Earth, so how does David Benito-Alifonso feel about his work?

'We don't feel like we're playing God. With our technology we are just helping the plants achieve their true potential. Is modern medicine playing God by extending our lifespan and not dying at 40? We don't think so, it's just helping us live longer and healthier.'

Desirable destination

Saving 8 per cent of our emissions by assuming 20 per cent increase in crop yield and using the saved area for carbon-sucking land uses such as forestry or wild meadowland.

How to get there

- Perfection and widespread adoption of enhanced photosynthesis and other yield-improving technologies.
- Proven safety and public acceptance of more biotechnology in agriculture such as nanotechnology or gene editing.

Fringe benefits

- More food in areas prone to poverty and hunger.
- More land available for wildlife and nature rather than farming.

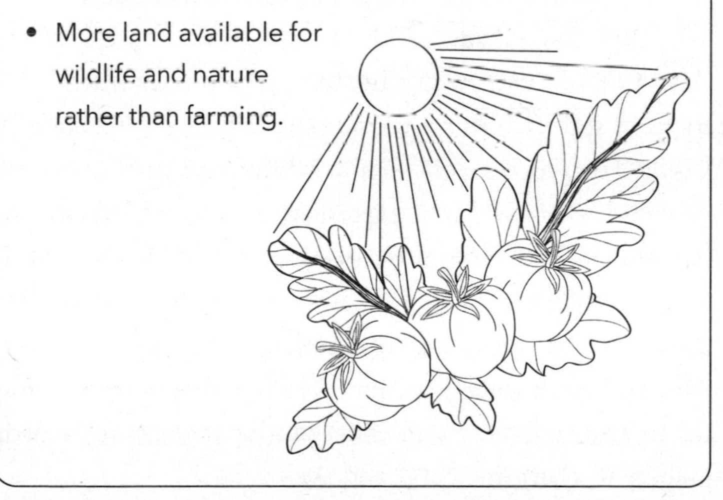

20
Rock Rescue

On the sands of Torloisk beach, in the Hebrides, I have drawn mazes, dug sandcastles, flown kites and run races. The grains are deep grey with occasional sparkles, worn off the punctuating black basalt piles along the coast. The dark rocks and sand both warm in the sun, so an incoming tide makes swimming on this beach a little less chilly than at its fancier golden neighbours. It's a place I love and have visited all my life but now I discover another reason to warm to it: basalt sand absorbs carbon dioxide.

That's enough fun on the beach, time for a chemistry class. CO_2 in moist air or dissolved in rainwater reacts with rocks or soils rich in magnesium, calcium or silica such as basalt. The process, called mineralisation, is similar to the chemistry described in Chapters 27 and 35. The result is a bicarbonate, magnesium bicarbonate or calcium bicarbonate, which collects on land or is washed into the sea, where the carbon is stored for hundreds of thousands of years. All that bicarbonate is also a dilute alkali helping combat ocean acidification, a side effect of global warming, which is currently harming coral and shellfish.

This mineralisation or weathering process is a natural phenomenon that already absorbs around a billion tonnes of carbon dioxide ever year from the atmosphere. Fans of 'enhanced weathering' think that we could possibly increase that tenfold.

The Leverhulme Centre for Climate Change Mitigation is suggesting spreading crushed rock on farmland to create a huge chemical 'sponge' to soak up more CO_2 from the atmosphere. Soil microbes and plant roots help to speed up the essential reactions. Geochemist Professor Rachael James is cautiously optimistic: 'Early results look promising in terms of extra carbon dioxide removal and also it tends to improve the fertility of the farmer's fields. But how you source the rock and get it to the field has to be done right. We don't want to dig lots more mines and quarries.'

To do enhanced weathering at scale will demand millions of acres of farmland and billions of tonnes of rock; without care, this could mean a huge and counterproductive fossil fuel demand. So, let's start with the source of the mineral. Rachael James says suitable rocks are widespread and frequently left over from other mining activities: 'Take diamond mining: you crush at least 1 tonne of rock in the hope of finding 1 gramme of diamond. Often storing mine-waste products on site becomes hazardous, as we saw in Brazil in 2015 when a dam holding back tailings from an iron ore mine collapsed killing 270 people. That material could have been used for enhanced weathering.'

Some rocks would still need grinding to a sufficiently fine powder and this would have to be done with low-carbon energy. But our history of digging stuff up from underground has left historic piles of potential raw material across much of the world. This matters because ground rock is heavy so you ideally want the field close to the quarry. Then you have the energy demand of spreading it on the fields themselves. Low-carbon trucks and farm machinery help the equation. Luckily, as Rachael James says, spreading these minerals can increase crop yields too.

On a hillside overlooking the River Tay in Scotland, I met Alec Brewster among his Aberdeen Angus grazing on the slopes. On closer inspection, between the blades of grass I could see granules like dark, chunky rock salt. This field has been showered with ground-up basalt and tests have shown it is beneficial for microbes that encourage plant growth and those involved in carbon sequestration. 'There is the potential to stabilise carbon with this product and that is in the forefront of humanity's mind right now – can we stall global warming. Farmers can be slow in changing but we are in a good place to front up some really big win–wins here.'

Its double impact stems from helping both crops and climate. Alec Brewster remembers his mother recommending spreading rock dust on the soil to grow great vegetables. Many rock types are rich in useful chemicals such as potassium or phosphorous or trace nutrients such as zinc and magnesium. This could reduce the demand for nitrogen-based fertilisers – a huge greenhouse gas contributor.

Other land uses such as forestry, palm oil plantations and even peatland restoration seem to benefit from some light rock dusting. Many farmers have used crushed limestone for years to reduce soil acidity; adding silicate rocks instead would do the same *and* capture carbon. Rachael James's own research is looking at trials in Malaysian Borneo, farms in the UK and in the corn belt of America.

With so many variable costs – source, transport, spreading – and so many different benefits – higher yield, better nutrition, less fertiliser – estimates for the price of removing carbon this way vary widely from US$39 to $480 per tonne. But economies come with scale and there is potentially vast

scale in this idea. Around 11 per cent of the world's land area is planted with crops and if two thirds of that were spread with the right rock it could absorb between 0.5 and 4 billion tonnes of CO_2. The top end is nearly 10 per cent of our current emissions. But at that scale we might have to start quarrying basalt or similar rocks specifically for the purpose and that would need 2–3 billion tonnes of stone every year and a business about one third the size of global coal mining today.

Let's get back to the beach and Project Vesta, a surf-based solution from an American team. Coastal enhanced weathering (CEW) uses the tumult of the waves to speed the critical chemical reaction. The Project Vesta team plan to spread olivine, a green volcanic rock made of magnesium silicate, on to frequently stormy beaches. The churning water will help grind the rock into finer particles and make it more available for the chemical reaction that absorbs carbon dioxide. The team also stress that it makes the resulting dissolved carbonates become readily available to build corals, shells and skeletons of marine organisms. The critical question is similar to other forms of enhanced weathering: 'Do the carbon-capturing benefits outweigh the costs, risks and energy input?' But their ambition is stunning and they cite the possibility of absorbing trillions of tonnes of CO_2, which could suck up all our emissions, by spreading green sand on 2 per cent of the world's most energetic seas. Such numbers rather stretch plausibility but maybe our idyllic seaside view will change from gold and blue to green and black.

Desirable destination

Absorb 4 per cent of our greenhouse gas emissions every year from 2040.

How to get there

Decarbonise: ensure that the quarrying, crushing, transport and spreading of the rock uses minimal fossil fuels.

Fertilise: improve research into the effect of rock dust on soils and spread the knowledge through farming.

Industrialise: enlarge and streamline the rock dust supply chain to make it available and cheap.

Fringe benefits

- Less fertiliser use.
- More nutritious food.
- Using quarry waste.

Society

21
Girls' School

All of the ways to cut climate change in this book also deliver other good outcomes, listed as the 'Fringe benefits' at the end of each chapter. Seagrass provides fish habitat; dry seeded rice uses less scarce water and making biochar could reduce the risk of forest fires. But this solution has so much *else* going for it that the carbon-cutting benefit almost feels on the fringe. Teaching girls has been termed 'the best idea ever'.

An educated woman has more of everything: more skill, more knowledge, more choice, more power, more money. Well, not quite everything: she tends to have fewer children and that helps the climate. It's estimated that ensuring all girls complete secondary education would shave close to 800 million off the world's population by 2050. As all our lives, to a greater or lesser degree, result in greenhouse gas emissions, then having fewer humans helps to avoid the climate crisis. Or, flipped around, humanity has a 'safe' level of CO_2 output and the more of us there are, the more likely we are to bust the boundary.

Esnath Divasoni comes from a family of six girls and one boy, brought up in the Zimbabwean village of Marondera, 120 km from the capital Harare. Her parents worked as both paid and subsistence farmers, and scarce cash was spent on food with little left for education. Secondary schooling costs around US$60 per year plus books, stationery and uniform. Aged 13, Esnath expected the classroom doors to be closed to her for good.

The overall global picture of girls' schooling is improving. In the past two decades, many countries have reached close to universal primary education regardless of gender. But it's secondary education that too many girls are missing. In low-income countries, only around one third of girls complete secondary schooling. Why? The primary reason is not simple prejudice – many poorer societies acknowledge

the benefit of well-taught women – but rather the unjust logic of circumstance.

In many poorer societies, uneducated women can still find a role as a bride and a mother whereas an uneducated man may well find it hard to get work and there may be no one else to support him. Schools are often a long way from home: boarding costs money and the journey harbours the risk of sexual assault for girls. Secondary schooling often coincides with puberty and taboos surrounding menstruation and limited access to costly sanitary products can be further obstacles. Put these together, when money is tight it's the boys who go to big school.

But Esnath Divasoni got lucky. She got a bursary from CAMFED – the Campaign for Female Education – a charity founded in Zimbabwe in 1993 and working across sub-Saharan Africa. It has supported around 4.5 million girls. With fees paid, uniform bought and some extra money for food, Esnath excelled, going on to university and now returning as an inspiring businesswoman. 'I am super-proud, being part of a movement that is changing the world and creating history by living it. Though it scares me to think what my life would have been without CAMFED.'

She doesn't have to look far for a clue as many of her friends dropped out of school aged 13 and now have five or six children: 'The only thing they know is giving birth. But sadly, having lots of children is not an achievement but rather a burden on you and the planet.' Esnath has one son, Adel Munashe, now nine years old.

The importance and urgency of educating girls is well recognised globally. Malala Yousafzai won the Nobel Peace Prize in 2014 for her campaigning work to keep girls in school – work which nearly cost her life in Pakistan when a gunman boarded the school bus and shot her in the head. The World Bank prioritises funding for female education as one of its recent reports said the lack of it was costing the global economy trillions of dollars in lost productivity. It is also central to big charitable funds such as the Bill and Melinda Gates Foundation. The campaigns and the cash are working: across the world girls' education rates are rising with countries such as Bangladesh increasing secondary school enrolment for girls from 39 per cent in the 1980s to 67 per cent in 2017.

Fiona Mavhinga, CAMFED's executive adviser, thinks success is a reason to go faster. She benefitted from the organisation's help as a child, though, even with CAMFED's support, she still had to sell groundnuts at the market to afford stationery, books and a uniform. She has gathered some strong statistics: in sub-Saharan Africa, she tells me, educated women have, on average, 3.1 children whereas, for those who drop out, it is around 6. 'Educated women have a choice of when or if to get married and how many children they want. And those resulting smaller families can afford schooling more easily, creating a virtuous circle. So, it is a game changer on all fronts. There is a direct link between female education and population growth and a direct link between population growth and climate change. Taking action around climate change means taking action on female education.'

But CAMFED's ambition around climate change goes beyond fewer babies. Young CAMFED alumni are acting as guides on climate-friendly farming. They are promoting the use of mulching – spreading dried leaves and grass on the ground to hold in precious moisture and lock more carbon in the soil. Elsewhere they have been making organic fertilisers or promoting the use of old plastic bottles for drip-feed irrigation. Esnath Divasoni used her success at school to study agriculture at university and has now started an insect farm rearing crickets, grasshoppers and mealworms.

Insects, especially crickets, are a familiar part of the African diet as they can be easily caught around harvest time. Esnath keeps them in recycled plastic water tanks and feeds them largely on food waste. They are a low-price, high-protein, low-carbon food. But she is not trying to corner the market, quite the opposite. She has trained 25 other farmers so far, mostly from her old school or local community. 'They are up to 50 per cent protein and farming them emits 25 times less carbon than beef. I love them and to eat one cricket meal a day would be marvellous.'

Desirable destination

Combining at least secondary education for all women and expanding access to family planning could cut the global population by 800 million by 2050, which could save around 5 per cent of greenhouse gas.

How to get there

Government policy: funding and promotion of girls' education as a priority. Poverty alleviation strategies.

Safety: both schooltime and the journey from home must be free of any threat of sexual violence or harassment.

Wellbeing: availability of washing facilities and period products at school.

Discrimination: tackling the belief among some communities that female education is unnecessary or even undesirable.

Contraception: adequate funding and availability of family planning and sexual health services from international aid agencies and national government.

Fringe benefits (more *core* than fringe)

Better lives: women with more knowledge, power and choice.

Richer lives: a World Bank study found that every year of secondary school education is correlated to an 18 per cent increase in a girl's future earning power.

Equality: A goal in itself.

22

Eye in the Sky

Emissions of carbon dioxide and other greenhouse gases have the potential to end our civilisation, so you might think we would have a fully accurate idea of where, when and in what quantity they are being emitted. Sadly not. This crucial information is patchy, frequently superficial and usually arrives long after the event. Most countries are self-reporting their data with little verification. Reliable, detailed and up-to-the-minute numbers would allow all those interested in cutting carbon to make better decisions, be they companies, investors, inventors, campaigners, governments or lawyers. What's honestly measurable becomes truly important, and clear information on carbon dioxide quantities is what TRACE (Tracking Real-time Atmospheric Carbon Emissions) is planning to deliver.

'You can't manage what you don't measure and we're going to do the measuring part.' So says Matt Gray, CEO of TransitionZero and one of the partners in the TRACE project: a coalition of climate-concerned non-profits, tech companies and the American former vice president and campaigner, Al Gore. They are combining satellite imagery, machine learning and private and public datasets to create a web portal for monitoring greenhouse gas pollution anywhere in the world: real time, public and independent.

Here is how it works. In Europe, the USA and Australasia, the climate-relevant data on most big emitters – from powers stations to factory farms – is accessed by TRACE alongside what you can 'see', such as the smoke, steam, heat plumes of these enterprises, from Earth-orbiting satellites. And don't forget to add in the weather chart. This so-called 'training data' is then fed into artificial intelligence software. The programme then deliberately ignores some of the information and makes predictions of likely outcomes. The system can check its own educated guesses against a full dataset and steadily get more accurate. Once the programme has become 'intelligent', you can apply it to places where you lack full data, predominantly the global south and less open political regimes. For example, if you know the coal inputs, steam patterns, heat signatures and carbon emissions of a number of American coal-fired power stations in all sorts of weather, your algorithm can reliably infer the carbon emissions of an Indian power station just from the satellite imagery. It's like an experienced boxer being able to predict the incoming punch because he's learned the meaning of an opponent's body language. He has seen that kind of thing before.

The remote sensing data comes from an existing cluster of satellites that are imaging every part of the Earth, every day, down to a resolution of 30 metres. The computational expertise and data gathering on land and sea come from project partners, and while the process of producing these numbers may seem obscure, we know the power of a reliable metric. Just look at body mass index, school league tables or insurance risks: these numbers take some calculation but can dominate our lives. In fact, the urgency of identifying climate change culprits has sparked its own mini-space race: in early 2021 an outfit called Carbon Mapper announced it would be 'locating, quantifying and tracking methane and CO_2 point source emissions from air and space' with the help of around 20 new satellites to be launched by the mid-2020s.

A good example of something reliably measured giving real climate benefits comes from the US Environmental Protection Agency's Continuous Emissions Monitoring System, which delivers data every half hour on greenhouse gases coming from power stations. Big companies, such as Google and Facebook, who want to cut their carbon footprint can now match their electricity demand to times of high renewables output: that's when their massive data centres can run 'hot'. Conversely, when the electricity is coming mainly from fossil fuels, customers can try to turn down their demand and buy offsets based on reliable carbon figures.

There are, of course, more contentious uses of strong evidence: catching cheats. Matt Gray of TRACE is a little bit coy about being seen as the forensic wing of a global

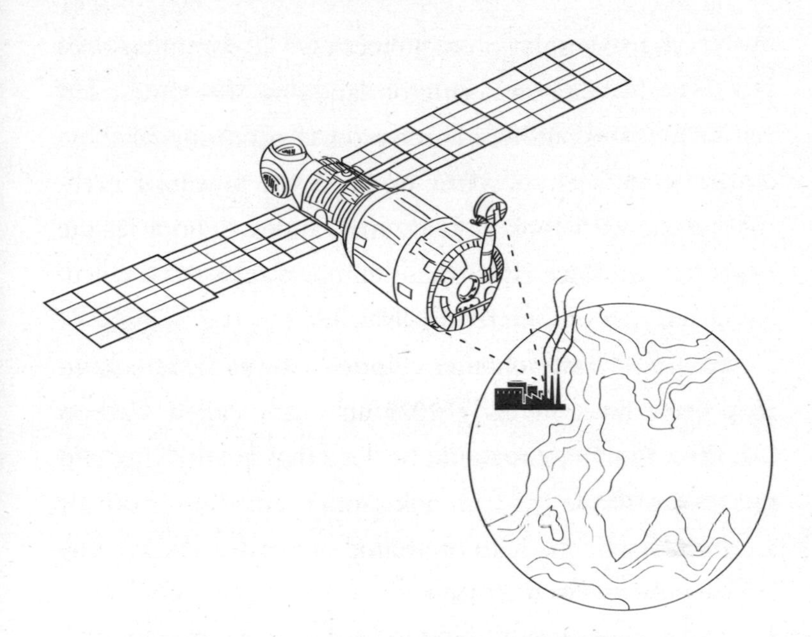

carbon patrol: 'We just make the data public and accurate, others decide how to use it.' But TRACE knows the power of making greenhouse gases 'visible'. Many countries have pledged to cut carbon by set amounts at international climate accords and this system will verify their claims and keep them accountable. 'It may identify data misreporting and falsification and this could be used by climate campaigners or lawyers.'

It's widely suspected by campaigners that some countries and companies already lie about their carbon score and, as this number becomes a greater badge of honour or shame, the temptation to do so will increase. Inescapable evidence will help keep them honest. In China, where climate data is not generally publicly available, Matt Gray expects the system to be very useful, but in an unexpected way. 'The situation in China is not what you think. The central government has committed to net-zero by 2060 but local government and state-owned enterprises tend to be motivated by different priorities like regional development or economic targets. So, the Chinese government could use TRACE to police their own provinces.'

Also, data from many low-income countries is poor due to poverty not secrecy. They simply lack the monitoring infrastructure because its expensive. TRACE can deliver the figures for free.

But TRACE isn't just keeping an eye on the obvious polluters such as heavy industry and power generation, it also has agriculture and shipping in its sights. For more than a decade, satellites have been imaging deforestation and land-use changes. Computer pattern recognition programs can then 'teach' the system what degraded land looks like and, combined with in-house expertise on regenerative farming, can then reliably estimate how much carbon a particular hectare is likely to be absorbing or emitting. Opening up this data will affect decisions made by governments, supermarkets and even individual grocery shoppers.

An eye in the sky is particularly useful for measuring emissions at sea. Within the TRACE coalition is OceanMind, whose background is in identifying illegal fishing. It analyses satellite photos, shipboard positioning beacons and radar data, with the help of artificial intelligence, to identify vessels doing the wrong thing in the wrong place. Mixing its software codes with the knowhow of others on greenhouse gas emissions has enabled its technology to deliver real-time and place monitoring of freighters' funnels.

At the heart of this climate change solution is the old adage 'Knowledge is power.' We do of course have another one, 'Money is power,' and Matt Gray thinks TRACE's biggest climate hit will come from combining the two: empowering the world of finance to withdraw cash from where it's harming the planet and invest where it's helping. In short, TRACE could enable money to do the right thing.

Desirable destination

Accurate, real-time and trusted data on emissions of CO_2 and other greenhouse gases enabling avoidance, reduction or even punishment. It is an enabling technology boosting carbon-cutting regulation, policy and investment.

How to get there

- Deployment of Earth observation satellites.
- Open access to reliable emissions data.
- Improving accuracy of data from more diffuse emissions such as farming or forestry.

Fringe benefits

Scientific advances and new jobs in satellite remote sensing and machine learning.

23
Carbon in the Dock

In scouring society for breakthroughs on climate change, we tend to focus on inventors or engineers, and maybe even educators or farmers, and all have a huge role to play. But let's swap the hard hat for a grey wig, the spanner for a gavel, and head to court.

'The law is an absolutely indispensable tool in the fight against climate change. Law is the way that a society captures what we believe in any moment in time. And when it comes to climate change, we are going to need to fundamentally change our behaviour – the way we run industry, transport, make electricity. And all of that has to be captured in rules ... and that's law.'

Those are the words of James Thornton, founder and CEO of the environmental law organisation ClientEarth. It has scored notable victories in the UK and mainland Europe on air pollution and climate change.

Thornton himself is American and believes his home country's experience with the civil rights movement in the second half of the twentieth century forged a mould for environmental law. In essence it is about using legal means to empower the oppressed: be it black communities or the natural world: 'If you were a person of colour and asserted your rights you got arrested. So, lawyers had to get involved to get you out of jail. Then they helped to write the new laws to give equal freedom to everybody. Some of the founders of the civil rights movement saw the environmental movement as equally controversial and got some of the best young lawyers involved from the beginning.'

So how can these keen minds use the law to cut carbon? By legally testing government action or inaction, the constitution itself, international treaties and human rights obligations. Companies can face legal challenges from shareholders over investment decisions or plaintiffs suing for damages. Lawyers can also be the driving force in shaping new legislation enshrining carbon cutting into law. All this adds up and, according to the Grantham Research Institute on Climate Change and the Environment, in 2021 there were 2,092 climate laws and policies around the world alongside 420 climate litigation cases.

One of ClientEarth's most influential successes to date has been prosecuting European governments over urban air quality. The European Union developed health-based limits on concentrations of pollutants such as nitrogen dioxide, sulphur dioxide and fine airborne particles. Many European cities broke these limits because governments lacked policies to tackle the problem. ClientEarth took them to court and victories in the UK, Germany, Italy, Slovakia, Belgium

and many more countries have resulted in clean air zones, diesel bans, hydrogen buses, electric taxis and cycle lanes. Although the cases were about air quality and carbon dioxide itself was never in the dock, reducing greenhouse gas emissions has been a massive collateral benefit.

One of the most important legal cases clearly focused on climate was brought by the Dutch NGO Urgenda against its own government for inadequate emissions reduction targets. The claim was based on the need to respect international obligations agreed in the Paris Agreement of 2015 and the Dutch state's obligation to respect the right to life and family life written into the European Convention on Human Rights. The court of appeal decided that science was sufficiently clear that climate change is an imminent, even current, threat to the rights of Dutch citizens; the Supreme Court agreed and the state was ordered to toughen up its emissions reduction strategy.

Court rulings to defend the climate do not happen solely in richer countries. Dr Joana Setzer, legal expert at the Grantham Research Institute, catalogues environmental legal action around the world and says there has been a huge increase since 2015, much of it in the global south as countries there have felt the impact of a changing climate and their younger, less entrenched, judicial systems can be more radical. In South Africa, a massive new government-backed coal-fired power plant was challenged in court by the charity Earthlife Africa. It won as the government failed to include climate change in its environmental review. In Pakistan, a farmer successfully sued the government for failing to enact its own climate change policy.

It's not just governments feeling the legal heat over climate change: companies are also in the dock. Back in the Netherlands, in May 2021 a landmark ruling ordered the oil giant Shell to cut its emissions by 45 per cent by 2030. Environmental charities such as Friends of the Earth and Greenpeace successfully argued that Shell was legally obliged to align its policies with the Paris Agreement. ClientEarth used company law to block a coal-fired power station in Poland by acquiring shares in the company and having an economic study conclude that generating electricity with coal was a poor long-term investment. James Thornton is proud of his small but potent portfolio: 'We sued the directors personally, saying they were ruining our €30 investment, they should be investing in renewable energy. We won the case in court and the company's stock-market value actually went up.'

Dr Setzer says cases are also brought against corporations as a public relations weapon – even if the company wins in the law court, they often lose in the court of public opinion. A striking example is *Lliuya v. RWE AG*: Saul Luciano Lliuya, a Peruvian farmer, against a giant German energy producer. He alleges that the greenhouse gases emitted by RWE over time have melted Andean glaciers threatening his nearby town with flooding. He is only seeking €21,000 in damages but the harm to the corporation of this media-friendly David and Goliath battle goes way beyond that. But it's not just PR: hearing legal debates over how much blame an energy company bears for melting glaciers on the other side of the world is very useful and a decision could set a significant precedent.

But perhaps lawyers could have their biggest impact away from the courtroom and in the corridors of power. They are intimately involved in shaping new climate policy, be it the European Climate Law, the US's Green New Deal and the UK's pioneering Climate Change Act. In Brazil, lawyers have got together as a resistance movement facing the less than climate-friendly administration of President Jair Bolsonaro. They have worked together to draft new laws declaring a climate emergency and proposing new emissions targets to deal with it. At the time of writing, these laws were being debated in parliament.

Of course, every court case has two sides and there are plenty of brilliant lawyers fervently protecting the fossil fuel companies and the governments failing to address the climate change challenge. But overall Dr Setzer thinks she can see a change: 'Most lawyers and judges are now part of the solution.'

Desirable destination

Robust climate change law with clear targets and penalties in every country.

How to get there

Publicity: campaign for climate change legislation.

Democracy: vote for political parties promising effective climate legislation.

Technology: low-carbon innovation allows politicians to be bold.

Fringe benefits

- Cleaner air.
- Protected nature.

24

Meat the Future

'You cannot claim to be an environmentalist if you are still eating meat, dairy and fish. I don't care how many degrees you've got or how many speeches you've made; you can't be if you don't walk the walk.'

A few years ago, this bald statement was made to me by Hollywood director, James Cameron. The man who made *Titanic*, *Avatar* and creator of the *Terminator* franchise. Cutting meat from your diet is probably the most divisive carbon-cutting solution; seen by some as a screamingly obvious no-brainer and others as an assault on tradition and freedom.

'This topic plays out as a culture war with extreme points of view at both ends. It makes it more difficult for the big players, like governments, to act,' says Simon Billing, who runs Eating Better, an organisation seeking to cut meat consumption in the UK by half by 2030. But their approach is long on carrot and short on stick. 'You can't go telling people what to eat. You need to create a food environment where the food that is better for the planet is desirable for the diner. It's much easier to attract people than to push them.'

Simon's organisation works with the government, shops, farmers and chefs to promote appealing veggie food and high-quality meat, meaning a better quality of life for the animal, a better taste on the fork and better health. Also, less meat consumed means less land required for agriculture as grazing animals are rather inefficient converters of grass or feed grains into pounds of flesh. 'A few years ago, I went to one of our most prestigious culinary colleges and when I asked about vegetarian training, they said that came in the "boiled and steamed" syllabus. Things had to change.'

The production of food, from field to fork, accounts for about a quarter of our greenhouse gas emissions and, with different foods having such different carbon footprints, what we eat is one of the few areas where individuals have considerable influence. I am not offering a detailed breakdown of the climate impact of different foodstuffs (there are whole books on that) but, in broad terms, someone who eats loads of red meat; imported, out-of-season veg; and cheese for dessert, all washed down with a glass of milk, is likely to be responsible for warming the planet more than a diner with an appetite for fruit, vegetables, cereals and pulses who finds them seasonally and locally. Most of us dwell somewhere in between.

Andy Jones likes hospital food. He's a trained chef who likes school food too, and what's in the prison canteen, because he runs an organisation called PS100, the trade body of public sector caterers, who between them serve about a million meals a day in the UK. In the early noughties he was deeply involved in the campaign to make school food healthier for pupils; he now wants to make all institutional food healthier for the planet. 'We are the guardians of the nation's health, we bring them into the world. Sadly, in hospice care, we escort them out. We educate them and we have a chance to lead people to a change which is good for them and our climate.'

In 2019 he set a challenge to public sector caterers to reduce meat consumption by 20 per cent and make the remainder higher quality: less but better. Andy Jones's grandad had a smallholding in Leicestershire, so he spent much of his childhood around farmers and his brother went on to run a beef farm in Canada. He is not a vegetarian himself but totally gets the need for a more plant-based diet.

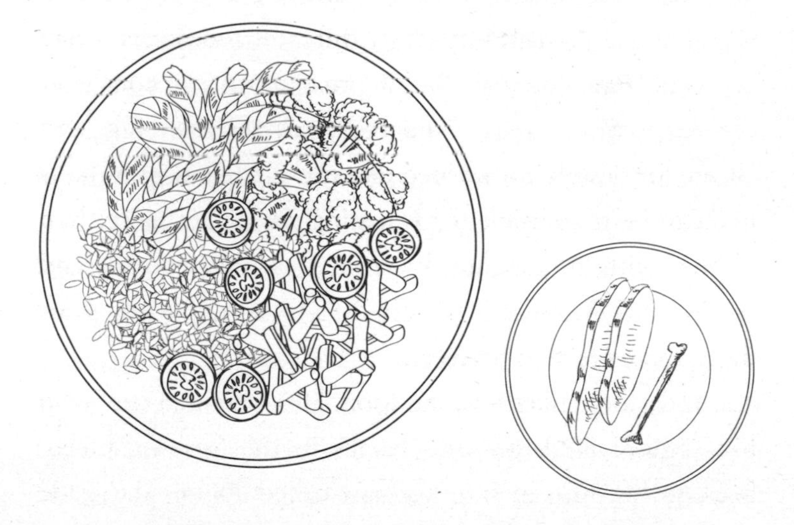

He bridles at the suggestion that this is just a middle-class fad, but does acknowledge that widely used institutions are a good way of reaching all sectors of society. Eating habits learned in school or experienced during a stay in hospital can go home with people. It can help to democratise the plant-based message.

His campaign seems to be working with nearly four-fifths of caterers reducing meat across their menus, and about the same proportion increasing proteins from beans and pulses. Just over half are using less pork and processed meat and, once again, a similar fraction are serving 'fake meat' alternatives.

This has been achieved with a mixture of pricing, nudging and better cooking. Portsmouth University offered a veggie loyalty card – 'Kale Yeah' – where your seventh vegetarian purchase was free. At Edinburgh Napier University, 'Meat Free Mondays' boosted plant-based meal sales by 3,000 across the year. Many care homes successfully encouraged more plant eating simply by putting the vegetarian options at the start of the menu. Nottingham University Hospitals Trust is launching a new standalone vegan menu with all ingredients sourced through local suppliers.

Simon Billing at Eat Better says the broadening appeal of plant-based food can be seen in the aisles of the supermarket giants – the veggie ready-meal section is mushrooming. But he says they could be doing so much more to use price, promotion and prominence to accelerate the trend. To get big changes, Simon says, we need to get the big players on board. That is being done successfully in decarbonising electricity with energy companies, government and inventors pulling together; we now need to see the same with food because emissions from farming and the food sector have remained stubbornly high. 'The net-zero target is a game changer because you can't ignore food emissions any longer, yet there are people at the top table [of government] who would still rather not talk about diet.'

Our decisions over food are a good example of how we respond to environmental concerns generally. Whether it is driving smaller cars, taking fewer flights or turning down our heating, a few of us will change our behaviour because it is the *right* thing to do, but most of us won't and moralising does little. In fact, Simon Billing thinks that, in order to take big strides towards a more climate-friendly diet, we need more pressure for change on the institutions and less at the dinner table: 'I don't want to get into a conversation about what you've got on your plate and what you eat. At least not from a climate sense. Food is about enjoying things together, a positive experience without guilt.'

Desirable destination

Halving the amount of meat and dairy we eat by 2050 to cut greenhouse gas emissions by 7 per cent.

How to get there

- Improving the quality, availability and desirability of vegetarian options.
- Greater animal welfare regulations to heighten livestock wellbeing and raise the price of meat, driving the shift to 'less but better'.
- Reliable carbon labelling to push more climate-friendly production methods through consumer choice.
- Better food and cookery education to spread skills on preparing tasty plant-based meals.

Fringe benefits

- Better human health from eating less processed meat.
- Higher welfare for animals.
- Less livestock means more land left for nature.

25

Indigenous Knowhow

Victor Steffensen is talking to me as he takes a break from fire-starting in the heart of Adelaide, South Australia. 'I'm surrounded 360 degrees by high-rise blocks, cars and flyovers and I've just put a match to this park.'

Victor is descended from Vikings on his father's side and the Tagalaka people from north Queensland on his mother's. This isn't arson, it is cool burning, a skill he learnt from his aboriginal forebears. 'Fire doesn't have to be threatening. Done right, it can protect life.'

Indigenous peoples make up just 4 per cent of the world's population but their territory covers nearly 25 per cent of the land and 80 per cent of the biodiversity. From depleted equatorial jungles to melting Arctic ice floes, their homes are often on the frontline of climate change but many reject being seen simply as symbolic victims and vehicles for rich-world guilt: their mastery of the natural world can help and they are fed up with being ignored.

Victor Steffensen believes engaging their wisdom would be a climate change solution: 'Aboriginal science has thousands of years of knowledge but Western science doesn't listen. It just ignores us and that is so frustrating. Western science is so young.'

Australia suffered devastating wildfires in 2019/20, with a scorched area a little bigger than England and Wales combined. The fires were both worsened by climate change, which triggered exceptional drought, and contributed to a warming planet through the massive carbon emissions. Victor Steffensen is leading a growing group that believes they could have been largely avoided if the country had followed aboriginal land management.

Burning at the right time and in the right place had always been a key part of both protecting the bush from bigger blazes and encouraging the growth of more plants and animals: reduce fuel and encourage food. The fires were contained and relatively cool, mainly fed by sparse grass. That left not only the trees and their canopy intact but also the swift passage of the fire just skimmed the soil, protecting future seeds and soil carbon. The fires were once such a feature of Australia that the explorer Captain James Cook commented on the smoke he could see from his ship.

Victor Steffensen learned how to 'read the land' by hearing from elders. He looks at landscape, trees, soil temperature and even certain flowers to gauge when and where to light the brush. This knowledge was honed over centuries to deliver a very sophisticated skillset that adapted vegetation to yield more plants and animals to feed the Aborigines and removed the threat of perilous wildfires.

They altered nature to thrive alongside it, which is the hunter-gatherer's equivalent of smart farming.

But in the last 200 years these skills have been either lost or actively banned. Many aboriginal groups were settled in National Parks, where burning was specifically outlawed. Invasive species of grass and trees, which are often more combustible, took hold in places. Homes and towns were built in a tinderbox.

'In 2019 there was no preparedness in the landscape,' says Victor Steffensen. 'Some places hadn't seen a fire for 40 years. There was the wrong vegetation, masses of fuel, a severe drought; it was a time bomb.'

When that 'bomb' went off and the wildfires raged it was noticed that often when the blaze met places managed by Aboriginal burning it petered out, they were protected. Since then, Victor and others with similar knowledge have been much in demand to help land managers and fire services plan better. But he says it's a slow process. 'It's hard because not enough people know how to read the land and those in power don't want to let go of the reins. But it boils down to following natural lore not Western law.'

Such phrases, though powerful, can come across as hippy cliché; all very lovely though not really relevant to a technologically driven planet of 7.5 billion people. But as we face the climate emergency, so many of the solutions, including many in this book, are about new ways of working in harmony with the land and sea. The climate science establishment calls them 'nature-based solutions' and reinforces the concept with conferences and research papers. Victor says we are just waking up. 'I do roll my eyes. The dash to "nature-based solutions" is welcome but so late. It validates everything indigenous voices have been saying pretty much forever.'

A big part of the problem is that climate science and indigenous knowledge come from completely different cultures and speak a different language. Diana Mastracci – a Venezuelan who studied anthropology in Scotland and lived for years with Siberian nomads – is trying to translate. It might help that she speaks five languages. She holds regular 'hackathons' with indigenous and Western software experts to devise apps that combine traditional wisdom with hi-tech resources, especially Earth observation satellites. 'It's important that the community choose how research looks and include proper validation and respect to the knowledge of elders.'

Such techniques are helping the Shuar people of Ecuador replant degraded forest, the Samburu tribe of northern Kenya to understand changing animal migration patterns and the Inupiat natives of Alaska to foresee shifting sea ice. In each case there was a two-way knowledge transfer. For instance, the Inupiat held their own sea-ice records going back hundreds of years from their 'whaling captains' who hunted bowhead whales, from kayaks and the floe edge. Sea ice has always been an essential source of food and materials so knowing the habits of the ice was a matter of life and death. Now this locally gathered data can be merged with the view from space and fed into scientific models.

But fostering such cooperation is not easy as in many places science was, and still is, a wing of colonial or commercial exploitation. In Alaska, Diana Mastracci says the local Inupiat have a nickname for scientists: 'The arctic squirrel: they only come in the summer, collect stuff without asking and vanish, leaving nothing for the community.'

While most Western researchers now are much more respectful, prejudice still exists as Diana witnessed herself in Alaska when she asked some scientists about their work – their response somewhat explained but was definitely not excused by her Hispanic complexion:

'They thought I was Inupiat and they told me I wouldn't understand anyway or talked to me very slowly like a child. I was really shocked but other Inupiat girls said, "That's normal, that's how they treat us." I was embarrassed because I felt that's how they must treat the local community.'

The simplest indigenous solution to climate change would be to give them the power to exclude all the mining, farming, logging and drilling from their one quarter of the world, if that is what they want.

Or, in the words of Victor Steffensen when asked what the authorities could do to heed the voice of indigenous communities: 'Jump in the passenger seat and let us drive for a change.'

Desirable destination

Protection of the quarter of land occupied by the world's indigenous peoples and applying their knowledge in land, fire and water management more widely. Impossible to put a figure on the impact.

How to get there

- Reform governments, international institutions and scientific bodies to give indigenous people a role in decision making.
- End the exploitation of the remaining homelands of indigenous groups from tropical forests to the Arctic.

Fringe benefits

- Indigenous culture and languages respected by, and/or protected from, industrialised society.
- Social justice.
- Protected semi-wild habitats rich in biodiversity.

26
At the Wrong Altar

'You can blame GDP – gross domestic product – for the climate crisis and the failure to act against that crisis, you can blame GDP for the nature crisis and you can blame GDP for the inequality that exists across the world. It's just become this beast which needs to be stopped in its tracks.'

So says Sophie Howe, the future generations commissioner for Wales. If you live in Wales, your children's fate is in her job description. She is one of a growing band of environmentalists and economists who now think our obsession with GDP is fuelling the drive to climate catastrophe.

You hear and see GDP figures all the time on the news, often given as a percentage rise or fall and taken as a proxy for a nation's success or failure. 'China's GDP increase approaching double figures' = country doing well versus 'Europe's GDP growth stagnant' = continent in trouble. Gross domestic product is defined in the *Oxford Reference Dictionary* as 'the total market value of all final goods and services produced within a country in a given period of time (usually a calendar year)'. It has become the accepted measure of economic activity and national vigour. Six months without GDP growth are officially termed 'recession': the shameful brand of national failure.

But for those with a care for our planet's nature and future it has two big problems. Firstly, imagine a mighty oak you climbed as a child that gives homes to bats, birds and bugs. It provides shade in summer and shelter in winter. It locks up carbon for centuries. But it does nothing for the country's GDP; until you cut it down. Then it's economic boom time with jobs for lumberjacks, joiners, furniture shops and wood fuel sellers. GDP ignores both the intrinsic value of nature (natural capital) and the human wellbeing that natural assets can deliver now and in the future – GDP will carry on growing until you cut down the last tree. GDP's blessings fall on the money and activity that nature's destruction can generate.

Secondly, GDP fuels an obsession with growth that is logically incompatible with reaching a balance with the world's natural systems. Viewing the whole planet as a finite human body, some have said the swelling economy should be considered a malign growth – a cancer. If GDP growth follows a similar trajectory in the next 30 years as it did in the last, our global economy will be three times bigger by 2050, making the zero-carbon target, which many countries have tied to that date, so much harder to reach. A country that has decided its citizens on average and country in general have *sufficient* no longer needs to grow its economy but, by the measure of GDP, it would be deemed a total failure.

The solution is a law or metric tough enough to withstand the GDP onslaught, and in Wales they might have found one. 'The Well-being of Future Generations Act' requires public bodies in Wales to think about the long-term impact of their decisions on the environment, climate, poverty and health. The woman driving that 'thinking', Sophie Howe, sports some silver bling with the word 'Chopsy' adorning her neckline. It's Welsh slang for feisty woman and she's earned it: 'The Act is a shield I can use to defend against the dragon of GDP.'

She went into battle against a £1.4 billion plan to improve a 20-km stretch of the M4 motorway in the county of Gwent. The traffic congestion there had been described as 'a foot on the neck of the Welsh economy'. Figures for potential GDP growth were marshalled as the main argument for its construction. But how did it measure up in terms of future generations? A question of particularly direct relevance when the borrowing required to pay the hefty bill would be paid back by those future generations. Traffic would increase, air pollution may worsen, carbon emissions would rise and the important Gwent Levels nature reserve would be harmed, whereas a fraction of that money could be spent on new train stations, new bus routes and greater provision for walking and cycling. All of these better protect the future health of both the people and the environment.

Howe won, the bypass was abandoned, and she believes the Act must take much of the credit: 'The more holistic approach promoted by the Act helps keep the beast in check.'

But is GDP growth really such a beast? Should we really be 'unfriending' something that has delivered health, wealth and happiness over the last century or so. It's notable that in the World Happiness Index, a metric that some claim should be valued higher than GDP, the top five countries – Finland, Denmark, Switzerland, Iceland and the Netherlands – are all pretty well off.

The respected economist Diane Coyle wrote a book in 2014 called *GDP: A Brief but Affectionate History*. I asked her if it was time for that friendship to end.

'It's fading, but GDP is linked to jobs and also the living standards improvements that come from innovation. This is why I'm not a degrowther. If GDP growth is zero, I can only invent vaccines by making people give up something else.'

As mentioned earlier, GDP stagnation or decline is a recession and the human consequences are real. People suffer from unemployment, lost income and lower self-esteem. So, there are two paths here. One is developing genuinely green growth that provides good jobs but not at the expense of diminishing our natural resources. Many of the chapters in this book have described the growth of carbon-cutting industries and the jobs they would provide as a 'fringe benefit' as economic growth is not the focus of *39 Ways...* but many politicians and economists see this 'green growth' as the core reason to back decarbonisation.

The other path is to develop robust metrics that quantify and celebrate environmental virtues. One could measure the green growth mentioned above. It won't be easy because the dividing line between 'good' and 'bad' industries will be hotly debated but it's vital because, as we know from society, what's measurable becomes important. Another measurement rapidly gaining support is the 'Natural Capital Asset Index' that calibrates the loss or growth of natural assets such as forests, fisheries, biodiversity or clean water.

But for any metric to matter it has to be given prominence by the media which, according to Diane Coyle, is loath to let go of the familiar GDP narrative: 'Journalists are complicit in the dominance of GDP. The media has an important role in giving the measurement of natural capital the same prominence.'

Sophie Howe is even stronger: 'The media are very culpable for the worship of GDP.'

So let's give the last word to the economics editor of BBC News, Faisal Islam, a man who often delivers the latest GDP figures.

'There is no doubt that the way we measure the economy has unintended consequences. But the consistency of GDP is useful for comparing different countries and different times. Natural Capital measures are not yet readily understood by the audience so reporting the changes quarter to quarter wouldn't mean much.'

As vibrant nature is so key to our survival, it is time to make what is important measurable, not the other way round.

Desirable destination

Equal prominence given to measures of GDP, natural capital and wellbeing indices.

How to get there

- Further work by economists and social scientists to make natural capital measurements more robust and standardised.
- Media reporting of natural capital and wellbeing indices.

Fringe benefits

Less incentive to destroy the natural world.

27
Unstuffed

About one quarter of all our greenhouse gas comes from creating things such as roads and off roaders, clothes and washing machines, buildings and books.

The stuff that contributes most to making the world stuffy is cement and steel, and we cover in Chapters 34 and 36 efforts to reduce their carbon footprint. But that will not happen quickly so could we just use less? Julian Allwood, engineering professor at the University of Cambridge and author of *Sustainable Materials – With Both Eyes Open*, is convinced we can live well on less stuff by designing things better, keeping them for longer and learning to share. 'Take cars: we could travel perfectly well, in comfort and safety in smaller cars. Thirty-five years ago, when I was a PhD student, my car weighed about 750 kg. Now the average car weighs nearly double that. We currently make three times our own bodyweight in new steel every year for every single person alive on the planet.'

This 'stufflation' happens because efficient production and artificially low energy prices mean that raw material, be it cement, steel or cotton, is relatively cheap: cheaper than investing in smart people to reduce its use. Julian Allwood says this enables manufacturers to be greedy with ingredients and lazy about design. He cites the overuse of steel-reinforced concrete in office buildings: 'Most new ones are designed with a [load-bearing] specification of 7 kilonewtons per square metre. That's the equivalent of seven grown men in not just 1 square metre but every square metre. Being that close to six other men ... I'm not even sure that is possible.'

Many buildings are held up by the almost universal steel I-beam and this can be made as strong with 30 per cent less steel if you make it thicker in the middle of the span than at the ends as the laws of physics determine the centre is under greater stress. But this requires designing each beam for each span, not just cutting a length off a uniform strip: more work, more thought. Steel food cans could be 30 per cent lighter if they weren't designed to be stacked up to 50 high in the warehouse and the ingredients were cooked at a lower pressure, as done with aluminium cans.

Longer lifespan is partly about designing greater durability into a bridge, building or bed but it is also about our attitude to it. Land Rover started building its classic boxy Series 3 design in 1948 and more than half of them are still around today, partly because of their corrosion-free aluminium bodywork but also because people love them. I have a mountain bike that no longer fits me or any member of my family but I keep it because it is so beautifully designed and will be useful for future generations. Longevity can also be extended by versatility. Plan your building so its use can be adapted in future: the physical equivalent of software upgrades in the same computer hardware.

Julian Allwood's third remedy for excess stuff, after design and longevity, is increasing intensity of use. 'We have three built spaces per person: the place we sleep, the place we work and the leisure place like shops, restaurants or cinemas. So, we have three times more built space than we need. There is a huge market opportunity in reconfigurable rooms or furniture to change use at different times of the day. Do we really need *more* office space, especially after the experience of homeworking during the pandemic lockdown? And then our 28 million cars in the UK are used on average for just four hours a week: we could be sharing them with an Airbnb model of trusted owners and users.'

In fact, this is happening. Across North America and Europe, people are intensifying the use of their vehicle with apps that let other people borrow them for a fee. The smartphone portal to the digital world is shaking up another environmentally punishing business: clothing and consumer goods. Through social media platforms and dedicated apps, 'pre-loved' has become a new term of affection. Market analysts expect the global second-hand market to more than double from $28 billion in 2019 to $64 billion in 2024.

Aisling Byrne, founder of the clothes-swapping platform Nu Wardrobe, is one of the entrepreneurs driving the trend. 'I was really into fast fashion growing up but I was in India in 2013 the same year as the Bangladesh [Rana Plaza] clothing factory disaster that killed more than a thousand people and that for me was a moment of collective shame – why had I never questioned it? My biggest frustration when I came back was trying to talk to people about it and making them feel guilty without any real solutions. I became the person that nobody wanted to talk to at the party.'

Blessed (or cursed) with this new awareness, Aisling began to look at the environmental impact of fashion too: unsustainable water demand, misuse of agrochemicals and a sizeable carbon footprint from farming, manufacture and transport. Her solution is a site where novelty is recycled and you don't need money. By offering a garment of your own to swap you get a token enabling you to acquire something yourself. Items aren't precisely valued but there is a

silver or gold token depending on an item's quality. The receiver pays the postage. The more you put forward from your own wardrobe, the more you can update it. Women between the ages of 16 and 35 are the principal market and Aisling agrees that they don't need all these wardrobe additions, but that's not the point: 'We have to meet people where they are. We can give them affordable variety so they can move away from fast fashion without sacrificing.'

Repairing, recycling, swapping and second-hand trading are booming across apps, social media platforms and eBay. This means avoiding both millions of items being binned and millions more being manufactured with the associated carbon emissions. And this isn't simply being driven by the urge to get new stuff. People seem to enjoy giving as much as receiving: like the toys in *Toy Story*, we seem to like the idea of an item we loved being treasured once again.

Professor Julian Allwood welcomes these changes but says we shouldn't overestimate the environmental power of our domestic choices as only a sixth of materials end up in the home. 'The whole literature of sustainable consumption has been focused on households, so there are lots of academic papers about washing machines, which is fun but not that important. At home we don't buy very much cement and steel, so the most important things we can do about materials are at work. When our boss starts talking about new construction projects, we should be saying, "Let's reconfigure what we've got or, if you must rebuild, design it to last 200 years, not 30."'

Desirable destination

Fewer new things made with energy-intensive processes. More jobs created in the swapping, sharing and repairing economy. An enabler of carbon saving but unquantifiable.

How to get there

- Higher fossil fuel prices to encourage economic use of raw materials.
- Changing building regulations to enable greater resource efficiency and product regulations to enable product repairs.
- Question the need for big transport infrastructure projects.
- Further digital innovation to allow product sharing.

Fringe benefits

- Less waste.
- Less destructive mining for raw materials.

Transport

28
Slippery Ships

'Hello, I'm a barnacle looking for my first home, somewhere I really feel secure and attached. A solid foundation is essential, a corner position would be great and I love being among my own kind. Luckily, I've just noticed a huge, newly arrived, solid surface with a welcoming coating of marine slime and a few pioneering settlers. Sorted.'

Understanding barnacle decision-making is one of the more unlikely academic pursuits helping to save the world from excessive carbon dioxide ... but it matters. The fuel used for shipping accounts for over 2 per cent of human-made carbon emissions and it is increasing steadily in line with rising trade. A smooth hull slips easily through the water, minimising fuel consumption, whereas being encrusted with marine lifeforms is a real drag. Figures from the International Maritime Organization suggest that so-called biofouling increases fuel use by up to a quarter, with all the carbon consequences.

For shipping's climate impact, cleanliness is next to saintliness according to Anna Yunnie from Plymouth Marine

Laboratory's Centre for Marine Biofouling and Corrosion: 'If you've got any increased roughness on the ship, it's going to increase the drag. If you imagine the power it needs just to push a ship through the water and then you add on a whole layer of barnacles, you're going to need so much more power. There can be a 12–55 per cent energy efficiency loss. That can add one to three million dollars to the fuel cost of a month-long voyage.'

And that means a lot of unnecessary carbon dioxide belching out of the funnel. To show me what we are talking about, Anna Yunnie heaves an encrusted buoy on to a jetty from the waters of Plymouth harbour. Its alien jumble of shapes and textures looks like a collection of all the lifeforms that failed the audition for Pixar's *Finding Nemo* for being too ugly. But that does not trouble Anna: 'Fouling organisms float my boat,' she declares before identifying each one.

'If you put anything into the water, it's going to get fouled. Within hours you get a conditioning layer and then you're going to get diatoms, bacteria and even marine fungi clinging on. These will lead to marine invertebrates and their larvae. And then you get the macrofouling like barnacles, sponges and oysters.'

Blistering barnacles, mussels and other adherents to the underside of vessels have been the bane of sailors' lives ever since we ventured out to sea. They come in thousands of guises: bacterial film can coat a hull within hours of launch, swiftly followed by orange cloak sea squirt spreading like melted candle wax; tube worms leave rock-hard calcium bonded to the hull; oysters and mussels can develop colonies eight deep, while stinging hydroids, like pink-fleshy palm trees, 10 cm long, also attach themselves to the hull.

For centuries, such fellow travellers were known to slow ships down and, in the case of the British Royal Navy, weaken their prowess in battle. Naval architects discovered that coating the wooden hull with copper kept the creatures at bay, leading to our phrase 'copper-bottomed' for high-quality assurance. Copper promotes corrosion in steel-hulled ships but is still a key ingredient in toxic anti-fouling paints; at some cost to marine life more widely as the layers slough off into the sea.

Anna Yunnie's expertise is in the species that hitch a ride on ship's bottoms and how to stop them, but lowering carbon emissions is only one motive. It's the potential fuel savings that interest operators as, overall, billions of dollars are wasted in powering fouled ships through the ocean.

Contemporary thinking on how to keep a clean-bottom hull broadly splits into three categories:

Paints and coatings: these are toxic to fouling organisms or so slippery that even the initial slime finds it hard to stick.

Physical deterrents in the hull skin itself: sound pulses in the audible and ultrasonic range can deter some organisms, especially barnacles, but there are concerns about collateral damage for other sea creatures that use sound to navigate or communicate. The chemical company Akzo-Nobel and the electronics giant, Philips, have combined to create a paint with embedded ultraviolet LED lights. The

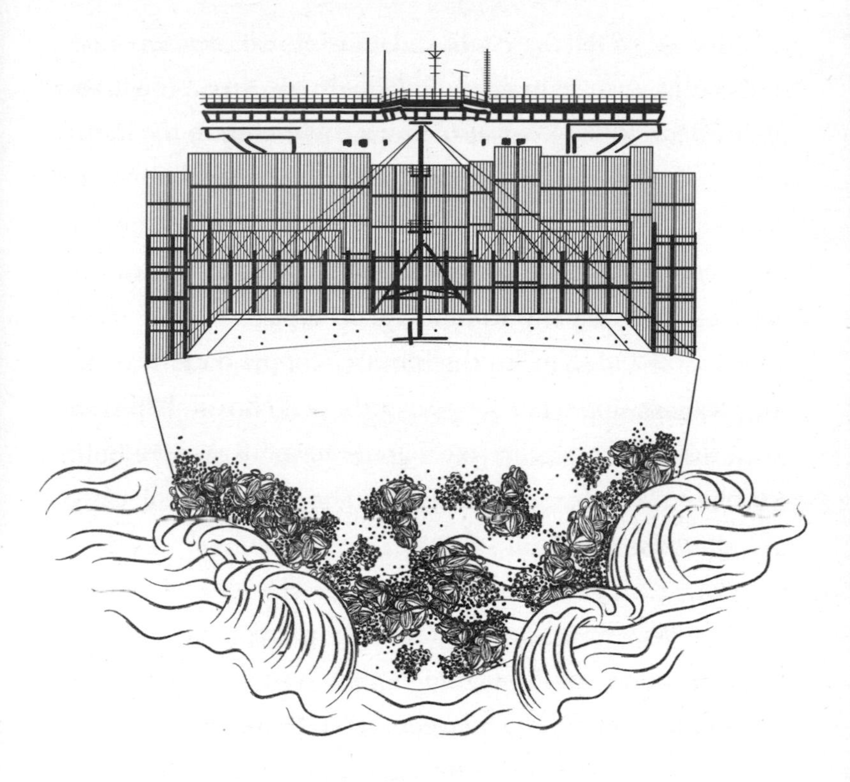

UV light provides safe and chemical-free sterilisation of the surfaces. The shape of the hull can also be designed to be less attractive to invaders: cavities should be avoided where possible and, if essential, should be accessible to cleaning. This is where the barnacle psychologists come in: advising how certain shapes and textures make the least desirable plot for a home-hunting crustacean.

Physical cleaning: this sounds like the most basic and in the past ships were scrubbed while being heeled over in the shallows but, now, the robots are taking over. Meet the HullSkater.

'Instead of relying on biocidal paint, we combine that with a proactive cleaning robot. It has three magnetic wheels each with a holding power of around 300 kilos. It has three cameras and brushes to remove early stages of sliming.' Geir Axel Oftedahl from Norwegian company Jotun is showing me one of his grooming robots, developed jointly with the engineering company Kongsberg, in a testing tank. It's a similar size and shape to the bonnet of a small car and the idea is that a ship will have one of these on board. When in port, the HullSkater can travel underwater all over the hull, a bit like a robotic vacuum cleaner. The key is to scrub early and often. Sea trials on 50 vessels are happening in 2021.

I suggest it's a bit like a tick-bird on the back of a rhino but Geir Axel Oftedahl has a better analogy: 'I think most of us brush our teeth morning and evening, we don't wait till they are full of calcareous deposits. It's as simple as that – good for the environment and good for the business of the ship owner.'

Desirable destination

Removing fouling from current and future shipping, saving 0.2–0.5 per cent of our greenhouse emissions. Eventually replacing ships' diesel engines with zero-carbon hydrogen and ammonia, saving 2 per cent of our emissions.

How to get there

- Deployment of the HullSkater and other hull-cleaning technologies, such as special paints and LED light skins.
- Increased fuel cost will make smooth hulls more attractive.
- Change the rules so emissions of a ship become the responsibility of the country where it is registered. At the moment, shipping in international waters doesn't count in any country's emissions total.

Fringe benefits

- Fewer invasive species: dirty hulls are the main way that unwanted 'alien' species travel the world and damage native ecosystems.
- Less air pollution: ship diesel, so-called bunker fuel, is relatively dirty and marine engine pollution is much less regulated than for road vehicles. In port cities ships can have a big negative impact on air quality.

29

Battery Power

Nothing drives change like the clarity of zero. Nowhere is that more true than in the recent decision by a number of European governments to permit *no* new petrol and diesel cars to be sold from 2030. It is the clearest of deadlines for the automotive industry but was only stated by the politicians when advancing technology made it feasible. The 100-year reign of the internal combustion engine will end in the next decade: deposed by the battery. Just as electricity shaped consumer goods in the twentieth century, mobile electricity underpins today's gadgets and resulting lifestyle changes – phones, tablets, power tools and now cars.

Transport overall accounts for around one fifth of our carbon emissions and a little under half of that comes from passenger cars, the rest from buses, trucks, ships and planes. So electrifying automobiles could take a big chunk out of climate change (assuming the generation source is renewable) but it's a change people have to *want*. If drivers don't come along for the ride, it won't happen. Many people love their cars and, confession time, I'm one of them. My car-ownership history goes Audi, Audi, Peugeot (a BBC company car), Subaru, Subaru. All the four I chose were unnecessarily rapid, consequently thirsty and I loved them.

My mum always said I had the 'broom-broom' disease, a common affliction in boys of the seventies, but thankfully some children, such as Isobel Sheldon, had the reverse reaction: 'My interest in engineering begun with a misconception. Aged eight, I was watching my dad mend our Ford Cortina estate. He told me how the engine worked by combining oxygen with petrol and I thought, "If we're consuming oxygen to make our cars go, is there going to be enough left to breathe?" That was very naive, but it got me thinking "there has to be a better way".'

The moment was a springboard for Isobel, who is now chief strategy officer for the leading battery maker British Volt. 'I was hungry for knowledge, I read most of the *Encyclopedia Britannica* between the ages of eight and eleven. I was always interested in different ways of moving people around that didn't involve burning things and creating pollution.'

Isobel went on to have a career in battery development including stints at the carmaker Toyota, the tech company Johnson Matthey and the UK Battery Industrialisation Centre. Throughout that time, using batteries to shift cars has had one big challenge: getting closer to the extraordinary potency of liquid fuel. The development of battery technology from lead acid to nickel cadmium, nickel metal hydride and now lithium ion has all been about squeezing more oomph into less space. They've done well but the bar is high. A top-end car battery today holds about 500 watt hours per litre, petrol holds around 10,000 watt hours per litre. Thankfully, electric motors are much more efficient than internal combustion engines so much more of the energy translates into motion but it explains why early electric models had such limited range.

The lithium ion (li-ion) battery is the same technology that powers your smartphone or laptop and it dominates the electric car market from the American Tesla to the European and Asian brands. It charges relatively quickly and can make that power swiftly available for lively acceleration; it's relatively safe, reliable and fairly long lasting. British Volt is building a 'giga-factory' outside Newcastle in the northeast of England to make lithium ion batteries for nearby carmakers. Isobel reckons that their energy capacity can still be improved by about 15 per cent but says there is another change coming that will improve the performance of electric cars. Until the early 2020s electric car sales were so low that the manufacturers (with the exception of Tesla) couldn't afford to design vehicles specifically for the weight and shape of electric motors or batteries. Now, given that approaching guillotine on fossil fuel vehicle sales, the design workshops are frantic.

'Most are going for a "skateboard and top hat" model,' says Isobel. 'A flat base of batteries, an engine at each end and a passenger compartment above. Bespoke designs can be more efficient, find space for more batteries and deliver greater range. I feel we are now at the tipping point I have been thinking about for 40 years. People don't think I am mad any more.'

Also, the increasing numbers of electric vehicles produced by the carmakers means they can now ask the battery makers for individual specification so a BMW drives differently from a Bentley or a Nissan. British Volt and most commercial battery developers are betting that the established strengths and future refinements mean that lithium-ion chemistry will power the automotive sector for the next ten years. But that means mining at least five times as much lithium while limiting the local environmental impact. One company in the UK is looking at extracting the metal from the deep groundwater under the western county of Cornwall.

In the longer term, other battery technologies may dethrone li-ion. Solid-state batteries that use a solid electrolyte (the conductive substance within a battery) rather than the liquid or polymer electrolytes used today are delivering around twice the energy density of li-ion in the lab. Similar leaps in performance are being attained in experiments with lithium sulphur batteries.

So, have all the improvement of electric cars, the hype and their green virtues cured this petrol-headed environmentalist of the 'broom-broom' disease? Partly. I have installed an electric charge point at home and, at the time of writing, I am leasing an electric car that does a realistic 350 km per charge. It's quiet, smooth, nippy and much cheaper to run than even a thrifty conventional car. In the two months and over 5,000 km of driving I have saved over half a tonne of CO_2. I have solar panels on my house and can set the charger to come on when the sun shines. But in truth I'm hedging my bets – driving it less but hanging on to the gas guzzler while waiting for its electric rival to get better and cheaper. Like a lot of people, I guess.

Desirable destination

Eliminating the 7 per cent of greenhouse gas emissions that come from cars.

How to get there

Law: more national governments need to enforce cut-off dates on the sale of petrol and diesel vehicles.

Tax: increase the duty on fossil fuels to make electric vehicles more attractive.

Technology: improve the range and lower the price of electric cars through innovation and mass production.

Charging infrastructure: ways to charge your battery need to be accessible, reliable and good value.

Fringe benefits

- Much improved urban air quality.
- Quieter cities and high streets as electric vehicles make less noise.
- Dual use of car batteries in grid energy storage.

30
Green Wings

'My philosophy is this: doing and showing is better than designing and dreaming.'

The words of Val Miftakhov, a pioneer of zero-emission flight and the boss of ZeroAvia. True to his word, he has flown the world's first hydrogen fuel cell-powered passenger aircraft and I do mean *he* has flown: alongside being an engineer and Silicon Valley entrepreneur, Val also pilots planes and helicopters. His father was an aircraft mechanic turned oil refinery engineer who played electronics with him from age six. Rivets run in the blood.

But increasingly something made that blood run cold – the realisation that his passion was killing the planet: 'Every time I went into these aircraft, I had a little sense of guilt.'

Aviation is currently responsible for 2.5 per cent of human-made greenhouse gases. Given its notoriety in the climate change debate, that may be lower than you expect but don't be lulled into a false sense of security. At average recent rates of growth and without low-carbon abatement that could triple by 2050; emissions in the upper atmosphere have a greater global warming impact; and for anyone who does fly it's likely to be a huge chunk of their personal carbon budget. The coronavirus pandemic of 2020/21 hugely reduced air travel and there are questions over how much business-related flying will resume given the imposed but successful adoption of online meetings. But we are not going to abandon the skies, so being up in the air without damaging the air is crucial. Hydrogen is the fuel to do it: its exhaust is water vapour, no more no less.

On board Zero-Avia's planes are tanks of hydrogen for fuel cells – mini power stations converting that H_2 into electricity (see Chapter 7). That wattage then drives propellers. A battery is also used to provide the extra punch needed for take-off. They plan to have a 19-seater plane operational by the late 2020s and a 100-seater capable of 1,000+ km by 2030 – a workhorse of mid-range aviation. But there are challenges and, as potential future passengers on hydrogen-powered planes, we care about what keeps us from a few thousand metres of free fall.

Have you ever looked at a jumbo jet or an Airbus A380 and thought, 'How on earth does that thing leave the ground?' The answer is a balance of power, weight and shape: hydrogen rewrites the rulebook.

First, facts to lift you up. Every kilo of hydrogen contains about three times more energy than a kilo of conventional jet fuel (kerosene). Fuel cells combined with electric motors are roughly twice as efficient as a mid-sized jet engine at turning that energy into propulsion. When you combine

those two, hydrogen gives you about six times more useful power than jet fuel.

Now, some facts to bring you down to earth. The hydrogen is stored as compressed gas in metal tanks. These are really strong – built to withstand crashes, bullets and fire – but they are really heavy. One kilo of hydrogen requires about 10 kilos of steel tank. So now, when you add the container to the equation, you have a power-to-weight ratio about half that of jet fuel. Efforts to improve these stats for hydrogen fuel systems are happening all the time – including testing the viability of having tanks of liquid hydrogen, as used in space rockets – but for now this limits their range. Long-haul hydrogen power may remain out of reach.

The other big hurdle is safety. It's not that hydrogen is intrinsically more dangerous than jet fuel – it's actually harder to ignite and vents rapidly upwards when released – but it is *new*. Safety protocols for conventional engines have been refined over more than a century of powered flight, and extremely rare risks have had decades to reveal themselves. ZeroAvia's fuel cells and gas tanks are bristling with sensors to detect any leaks and many components have backups. An intriguing safety-related challenge is potential overheating on take-off. Hydrogen fuel cells run best at about 80 to 90°C (176 to 194°F), which is far lower than jet engines but uncomfortably close to the 40 to 50°C (104 to 122°F) you could encounter on a tarmac runway under the tropical sun. Keeping cool is less of a problem once cruising in chilly high-altitude air but cooling tech could be needed at ground level.

ZeroAvia is an engine designer not an aircraft manufacturer and, in keeping with its philosophy of delivering something tangible soon, its systems are designed to fit within the airframe of existing planes. In the long term, aircraft shape could evolve to better suit bulkier hydrogen fuel cell propulsion. Further evidence of the practicality of the company's approach is the development of the Hydrogen Airport Refuelling Ecosystem at their test base at the UK's Cranfield Airport. Existing commercial airports have a complex network for delivering essential jet fuel to on-board tanks, and ZeroAvia is recreating this for hydrogen. It has partnered with another hydrogen pioneer from the other end of the UK – the European Marine Energy Centre, based in the northern Scottish Orkney Islands – which uses electrolysis to convert over-abundant wind energy into transportable compressed hydrogen for ships and trucks. Hydrogen fuel only makes climate sense if no carbon is emitted to produce it.

The ZeroAvia team say that there is no new 'fundamental science' needed to make commercial hydrogen flight a reality and claim electric motors should be cheaper to maintain than jet engines. But, as we have seen above, stripping the carbon out of flying requires considerable practical and structural changes to the aviation business. And there may be an even greater challenge: throughout our history transport advances have followed the mantra of 'further, faster, cheaper'. It's embedded in the aspirations of tourism and the economics of trade. Curbing those appetites in favour of 'cleaner, shorter, slower' flights will take considerable political courage. The technological fix that might avoid that uncomfortable choice is ammonia (NH_3)-fuelled planes. Liquid ammonia can be carried in tanks like kerosene and adapted jet engines could burn a combination of ammonia and hydrogen. The only exhaust is nitrogen and water vapour, but ammonia is toxic and explosive – these ideas are only on the drawing board.

A more plausible option for coming decades and for long-haul flights, which cause the bulk of emissions, is Sustainable Aviation Fuel or SAF. This is basically jet fuel without using fossil fuel, and the necessary hydrocarbon can come from a variety of sources, such as plant biofuel or waste fats. The most climate-friendly SAF is 'E-fuel', made by combining hydrogen electrolysed from water with CO_2 drawn from the air, but this would be very expensive.

On 29 April 2021, ZeroAvia's six-seater test plane came down in a field, distorting the fuselage and ripping off a wing. No one was seriously hurt and the hydrogen system remained intact. The problem occurred as they were switching from combined battery and fuel cell power to fuel cell only while at high speed. Val Miftakhov says it's a setback for the programme and upsetting for the team but insists incidents are in 'the nature of the beast' when you are testing new things and some problems can only be revealed under the genuine stress of flying. 'I'm absolutely confident in the future of hydrogen fuel cell flight. We are more confident than before the incident because we exposed one of the things we needed to work on and fixed it.' The programme for developing the 19-seater goes on but the accident confirms their decision to keep one of the two engines of that test aircraft running on conventional fuel in the first stage of testing, before proceeding to a dual-engine configuration for certification after that. In something as safety-critical as flight it's hard to break from the carbon-belching devil we know.

Desirable destination

Introduce hydrogen-powered planes in journeys up to 2000 km. Such flights account for about half the fuel demand of aviation currently so this could reduce our emissions by 1.25 per cent assuming no overall growth in aviation. Long haul would need as yet untested ammonia engines or synthetic sustainable fuels.

How to get there

Policy: government incentives to tax polluting flight either through increased fuel duties or a frequent flyer tax that raises the levy per flight as more plane journeys are made each year.

Investment: scaling up the research and development spend from the current hundreds of millions of pounds to billions of pounds.

Customer choice: business or leisure travellers choosing to take the time, cost or restricted-range penalty of greener flights.

Fringe benefits

- Less land take for airport expansion.
- Better air quality around airports.

Buildings and Industry

31
Wood for Good

In a cramped workshop off a basement corridor the air is filled with the smell of hot sawdust and the sound of splintering wood. As the pressure builds, split-cracks echo off the walls and the beam finally surrenders, but not before proving stronger than steel.

This is the lair of Michael Ramage who heads up the Centre for Natural Material Innovation at the University of Cambridge. He is a man who believes we can use wood instead of concrete and steel in most of our new buildings. This change alone could cut total global carbon emissions by about 4 per cent. He has just bent a steel bar with less than 1 tonne of force from his hydraulic press and is now demonstrating how his newly sawn length of timber, weighing the same as the steel, can withstand at least double the load.

The core logic is clear and persuasive. As trees grow, they capture carbon from the atmosphere. Once felled, that carbon can be stored as building material. Those beams, floors and walls also replace carbon-intensive steel and cement production. The forests are replanted to capture more carbon. And the tastiest statistic: every seven seconds the sustainable forests of Europe yield enough wood to build a four-person family home.

There is also a key innovation driving the new age of wood: cross-laminated timber (CLT). In the sawmills of Alpine Austria in the late 1990s, they started gluing together thick layers of sawn timber at right angles so the grain runs at 90 degrees to the layer beneath. This made, in effect, chunky plywood slabs, which can become a variety of shapes and sizes. These can be used for floors and walls, made offsite, cheap to assemble and with inbuilt insulation properties. Choose your woods and glues more carefully and you can create amazingly strong beams and pillars. The race upwards began and now 'plyscrapers' are competing to graze the heavens ever higher. A 53-metre student hall at the University of British Columbia in Vancouver, the 84-metre and 24-storey HoHo apartment block in Vienna and, just topping it by a metre, the Mjösa Tower in Brumunddal, Norway.

According to Michael Ramage, it is when building high that the amazing strength-to-weight ratio of wood really delivers. Many concrete and steel towers are limited by the sheer weight of the building above: lighter construction equals more potential storeys. And this weight loss delivers an added attraction in cities such as London undercut by underground trains. Much valuable real estate is above tube stations or near surface tunnels and you can't drive massive foundations down through them. A lightweight building is the developer's dream.

Plyscrapers catch the eye and, just like the original steel skyscrapers of the 1930s, are helpful to advertise the potential of the new material. But the real future for wood in general and CLT in particular is in normal buildings: homes, schools, warehouses and office blocks. I visited some CLT flats under construction in Hackney, east London, where the floors, walls and even lift shaft were being made of giant slabs of plywood. The pervasive pine scent and the relative quiet of the more sound-absorbent wood made it quite unlike the harsh ambience of typical building sites. Research suggests the lived experience amplifies this feeling: timber buildings benefit wellbeing according to an Austrian study of a wooden school where learning improved thanks to lower stress and heartrate. Michael Ramage's team in Cambridge and many other architects around the world are working on timber designs for low-cost housing and public buildings.

The Great Fire of London in 1666 started in a Pudding Lane bakery and ripped through the wooden city. A wood-burning stove is made of cast iron. The logs burn, the metal box does not. Surely wood is a greater fire risk? Tests in the USA have shown that CLT buildings can pass all the modern fire-safety regulations. One factor is the natural tendency of wood to form a charcoal layer on its surface, which protects the wood beneath; if you combine this with heat-resistant glues and, in some cases, claddings

such as plasterboard you can match or exceed the fire resilience of conventional buildings. And if you have ticked the boxes on your house insurance form for seismic or warfare risk then a timber home is definitely for you. The flexibility of modern CLT buildings makes them nearly earthquake proof and the US Department of Defense tests have shown them to have excellent blast resistance. You can waste time on YouTube watching swaying but sound timber structures on artificial earthquake plates or slow-motion explosions bow wooden walls.

Less than a mile from Michael Ramage's workshop beside the University Engineering Department in Cambridge is King's College Chapel, home of the famous Christmas carol service. Built in the late fifteenth and early sixteenth centuries, its intricate rood screen and vaulted oak roof were living trees in medieval Britain. They absorbed carbon dioxide over hundreds of years before meeting the Tudor axe around 1500. That carbon has remained captured ever since in impudent carvings and load-bearing beams: forms both useful and beautiful. If we use wood in our new buildings with such determination and creativity it is much more likely that our structures will be around for the next 500 years, along with a civilisation to admire them.

So why don't we dump the old ways and go back to the trees? Michael Ramage says CLT is relatively new and construction is a notoriously conservative and risk-averse industry. Habits are hard to change, especially among the volume housebuilders churning out brick boxes. What is needed is a carbon price that makes steel and cement less attractive, and an extension of government building regulations to cover not just the energy performance of a house when it is built but also the carbon embedded in construction.

Desirable destination

Save 4 per cent of our emissions by 2050 by insisting that the vast majority of new buildings are constructed mainly of wood, not concrete, bricks or steel, and being sure at the end of life the timber is recycled or burned as biofuel.

How to get there

Education: spreading knowledge of the potential of both new engineered timber and relearning old wood construction techniques.

Innovation: continued development of wood-based materials to improve strength, versatility and fire safety.

Regulation: building standards are already very tightly regulated so write in stipulations for wood. In France from 2022, every new public building must be at least 50 per cent wood or other organic material.

Fringe benefits

Dwelling in wooden buildings is thought to boost wellbeing – improving learning in schoolchildren and lowering blood pressure in adults.

32

Keep Warm to Carry On

Creating and controlling fire is probably humanity's greatest single feat of ingenuity: enabling us to cook, keep warm and make things. But the resulting carbon emissions are on track to be our undoing. Keeping our buildings cosy and our water hot still uses nearly one quarter of our energy and most of it is fossil fuelled. We need to reduce this and we can. For new buildings, it's all about regulation: the correct government policy to ensure what we are putting up today has exceptionally low energy demands. But for existing buildings – where most of us will live and work for decades to come – it's all about the insulation.

Insulation – the Cinderella of the low-carbon world, left behind by the darlings of green energy (heat pumps, solar panels and electric cars) – takes the spot-lit dancefloor. For at least 30 years, the need to improve home energy efficiency has been well known, but progress has been painfully slow and recently been undermined by growth in the average size of our homes. There are three main problems: it's expensive, dull and complex. Energy prices are too low. Not a popular statement but what I mean is 'too low' to make energy saving a no-brainer. It is often cheaper, even over a timescale of decades, to turn the gas up and let much of the heat escape rather than properly lag your house. Insulation is boring – the Tesla in the drive is more likely to stir the adrenaline or trigger an envious glance from a neighbour than the fibreglass in your loft. Improving energy efficiency is complicated with many different options and trades all incurring domestic disruption. So, three cheers for a Dutch company promising home insulation to zero-carbon standard that is simple, desirable and affordable.

'Upgrading the energy system of your home should be as easy and attractive as redoing your kitchen. Something you aspire to do. If you can't crack this, there is no route

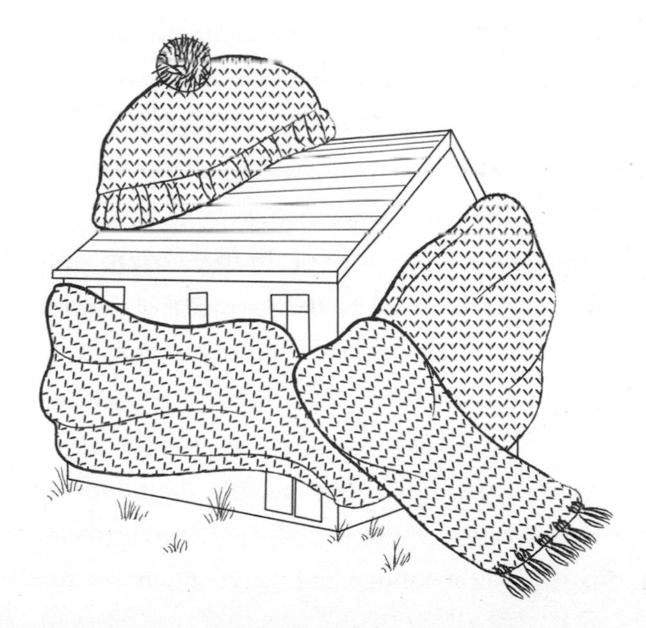

to scale. It is not realistic to think the majority of people will plough through the current domestic energy-efficiency swamp. And, by the way, there is usually no economic case for upgrading your kitchen but people will pay thousands for it because they think it will improve their lifestyle. We need to get there with zero-carbon homes.' This is the mantra of Ron van Erck, co-founder of Dutch-based Energiesprong. Most products sold in huge volumes, such as cars, phones or laptops, combine industrial construction with great functionality to create something desirable and affordable: Energiesprong aims to do the same with energy efficiency. It offers a zero-energy refit installed within a week and paid for over years in place of your – now non-existent – energy bills.

In practice, this means changing the way your energy is produced and how it is conserved. First, a 3D computer model of your home is constructed with scanners – drone-mounted where necessary. The roof is replaced with a top skin of solar panels and an underside of top-quality insulation. The external walls are clad with an insulated shell. An air-source heat pump is fitted in place of the gas boiler and ventilation is designed to both avoid damp and recover heat generated in the house by cooking, gadgets or just bodily warmth. All of these elements are designed and built offsite as such prefabrication means installation often takes less than a week. You get a comfortable, draught-free home with a smart new makeover and near zero carbon emissions. In the Netherlands, Energiesprong has completed these retrofits on more than 5,000 houses and 100,000 are in the pipeline across Europe.

But how it finances the makeover is as innovative as how it is made. Rather than paying for the work immediately after it is done, the company and the customer negotiate a fee for the 'Energy Service Plan' paid off over the next 25–30 years. Precise annual costs vary but typically it will be similar to the previous energy bill. You benefit from a more comfortable home and the planet benefits from zero emissions. Early adopters have tended to be associations or social housing providers who own large housing stocks of similar construction, which delivers economies of scale and they also understand long-term finance. But Ron van Erck says the idea works for private homes too as the contract to pay off the service plan stays with the *house*, not the owner. 'If you buy a property already improved by Energiesprong then you agree to take on the charge for the ongoing Energy Service Plan. It's actually less risky than buying a house with unknown energy bills.'

Ron van Erck thinks another way to help spread energy-saving innovations is to catch the moments of big expenditure: 'When people are buying houses or moving in, they often spend large sums much more readily and add it to the mortgage. It would be stupid to miss that chance to cut future energy bills and cut carbon.'

In a typical winter, across Europe, 50–75 per cent of total energy demand is used to keep buildings warm and it's nearly all fossil fuelled. Replacing that with renewable electricity is unrealistic and hydrogen heating is a way off yet. So, we need to cut that demand by copying success in other sectors. Recent, low-carbon, victories have been won where we have managed to industrialise the solution: solar, wind and, increasingly, electric cars being obvious examples. We need to do the same for domestic energy efficiency.

And while you wait, put on a jersey.

Desirable destination

In Europe and the USA, home energy consumption is responsible for 15–20 per cent of greenhouse gases and three quarters of that arises from heating and hot water. Halve that by 2040.

How to get there

Policy: subsidise the provision of insulation and 'deep retrofits', such as those by Energiesprong, that take homes close to carbon neutral.

Regulation: new building standards should insist on very high energy-efficiency standards.

Skills: a training revolution paid for by government and industry so builders, plumbers and electricians can fit and maintain low-energy materials and technology.

Energy bills: allow fossil fuel bills to rise for most customers to encourage insulation while compensating those in energy poverty through the welfare system.

Fringe benefits

- More comfortable, draught-free homes.
- Job creation for skilled builders and thermal engineers.

33 Carbon Dioxide = Sewage

We use water in our everyday lives and, after it has gone down the drain, we have a sewage system for its safe treatment.

We buy stuff in our everyday lives and, when we throw away what remains, we have a refuse collection for its safe disposal.

We use fossil fuels in our everyday lives and, after we use it and carbon dioxide has gone up the chimney, we have no system for safe disposal and it's slowly killing us.

This is the strange and sad truth of climate change. CO_2 is the sewage of oil and gas, the hazardous waste product from burning stuff. Yet the companies supplying the fuel do little or nothing to clean up the aftermath. It's time they were obliged to and we could all pay for it. That is the persuasive concept behind a less than grabby title: the 'Carbon Takeback Obligation' or CTBO.

Myles Allen is a climate scientist. In truth, that's a bit like calling Lionel Messi 'a footballer'. Myles Allen is head of the Climate Dynamics Group in the University of Oxford's Department of Physics and also professor of geosystem science in the School of Geography. He's a regular author of reports for the Intergovernmental Panel on Climate Change. And he is utterly convinced about the urgency and centrality of CTBO to solving climate change. So much so, that when I first explain the premise of this book, he asks, not entirely in jest: 'So, what will you need the other 38 for?'

Human-made climate change, the existential threat hanging over us all, is caused by too much carbon dioxide entering our atmosphere. The vast majority of that – 85 per cent – comes from burning fossil fuels. Our lives depend on stopping this by 2050, at the very latest, and there are theoretically two ways to do it. The first is to stop using fossil fuels, the second is to stop the carbon dioxide entering the atmosphere.

Option one is improbable. Despite the rapid advances in renewables, over 80 per cent of our energy globally comes from fossil fuels. They are still utterly dominant in shipping, aviation, heavy road haulage, heating and the manufacture of cement, steel and fertiliser. With electricity generation, though we are seeing huge advances in solar and wind, many countries will still use coal, oil and gas to keep the lights on for decades. Stopping fossil fuel use would demand not only giant technological leaps but painful sacrifices too: lifestyle changes likely to prove highly unpopular in Western democracies (so probably won't happen) and denying development to poorer countries, which will seem unjust (so probably won't happen).

Myles Allen: 'The main thing to realise is that we must stop climate change before the world stops using fossil fuels. The fantasy that we will simply stop climate change by simply banning fossil fuels or finding some alternative that's so cheap it's not worth extracting them is not going to happen fast enough.'

His answer for fossil fuels is to enjoy the power but end the pollution. One tonne of CO_2 must be safely and permanently removed from the atmosphere for every tonne created and this means carbon capture and storage (CCS).

Carbon capture and storage has three essential elements. The CO_2 is captured from the chimney as the fuel is burnt. This technology exists but is much more cost effective in big, highly concentrated sources such as power stations or cement makers. Then the CO_2 must be compressed into a liquid and piped to the storage facility. Storing CO_2 is like reversing the process of oil and gas extraction: it is pumped back underground and kept in the same rock formations that held the fossil fuels for thousands of years. Even the same offshore rigs that pumped the oil in the first place could be used to put the CO_2 back.

This striking idea comes from Margriet Kuijper, who spent nearly 30 years with Shell before running her own consultancy focusing on carbon storage: 'There are at least 20 large-scale projects worldwide storing CO_2 in that way worldwide. We know how to do it. Technology is not the issue.' But less than 0.1 per cent of CO_2 emissions are stored right now. 'The showstopper so far has been the business

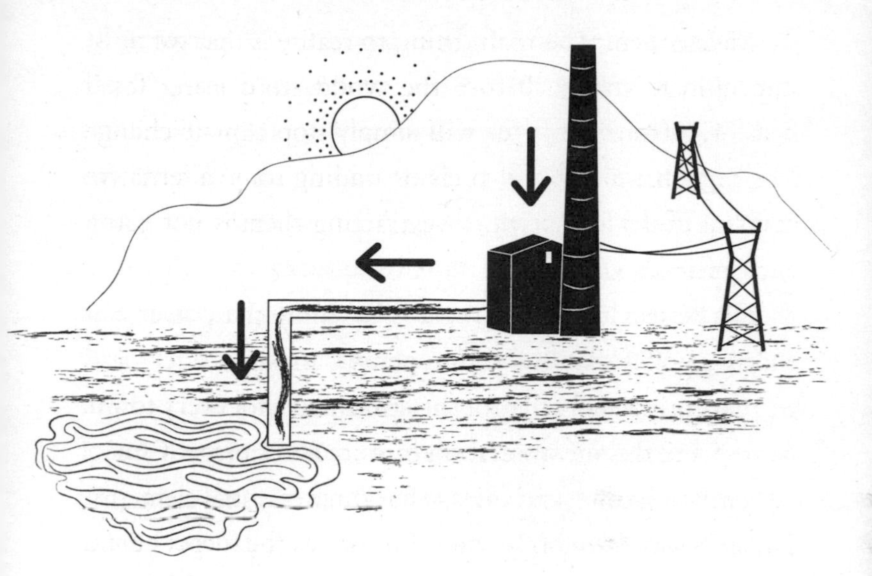

model. As now you can dump CO_2 in the air for free, why would you bother capturing it and storing it underground? That is always more expensive. Centuries ago, we just threw waste from the village over the fence, then we decided to clean it up. It costs money but it is important.'

Which brings us back to the sewage analogy at the beginning of this chapter. Imagine the Victorians looking at all the people dying of waterborne diseases such as cholera or dysentery and saying, 'No, let's not bother with a decent network of pipes and tunnels to remove and treat foul water. It'll cost a bit and looks quite complicated.' That is what we are doing today as the Arctic shrinks, sea levels rise, storms rage and crops fail.

Supporters of CTBO say the oil and gas companies or importers must be obliged to pay for storing a steadily increasing proportion of their product: 1–2 per cent by 2025, 10 per cent by 2030, 50 per cent by 2040 and 100 per cent by 2050. Those firms will initially squeal as they don't like regulation and are currently dumping pollution for free but in the long term, Myles Allen believes, they know their product will become more controversial and this gives them a better chance of survival – by cleaning up their mess they'll gain a social licence to operate.

Extra costs will fall on consumers, but they will be light at first. Even if we went straight to storing 10 per cent of CO_2 this is likely to add less that one penny or cent to the pump price of petrol. By the time we reach 100 per cent it would add a noticeable amount to the price of fossil fuel (it's impossible to say how much as economies of scale may drive CCS costs down) but that is part of the point – it's *meant* to make pollution-free energy such as wind, solar or hydrogen more attractive by comparison. It would be a big industry too – but so is the waste and water treatment industry and we don't question whether we need it.

A viable CCS system couldn't work with all the fossil fuel users that exist today: it doesn't make sense to have a carbon dioxide capture unit on every gas central heating boiler, let alone the pipework to remove it. So, heating in homes and offices would run on electricity or hydrogen from CCS-equipped power stations or hydrogen production plants. Natural gas would be a good raw material for both as burning it emits less local air pollution than oil or coal and, as its chemical formula is CH_4, the gas can be separated into four parts hydrogen to one part carbon: hydrogen would be piped out as a clean fuel (see Chapter 7), and carbon piped back below ground.

Margriet Kuijper says CO_2 was an unseen peril for centuries, a stealth threat. But now our knowledge has unmasked its true danger, we need to bury it. 'If CO_2 had a foul smell, we would have done something about it years ago. It has taken temperatures clearly rising, glaciers melting and wildfires raging for us to realise CO_2 isn't so innocent.'

I ended my conversations with Margriet and Myles with the same question: 'Can we reach zero carbon by 2050 without CCS?'

'No' times two.

Desirable destination

All large industrial sources of carbon dioxide, such as chemical works, hydrogen producers, steel furnaces, cement plants and high-energy manufacturing, fitted with carbon capture and combined with a massive increase in current geological storage facilities could cut greenhouse gas by 50 per cent by 2050.

How to get there

Policy: international agreement to place a steadily increasing Carbon Takeback Obligation on fossil fuel companies.

Technology: massive construction and rollout of engineering for the capture, transport and safe storage of carbon dioxide.

Acceptability: domestic and industrial customers have to be willing to payer higher bills for fossil fuels to cover the costs of the safe disposal of carbon dioxide.

Flexibility: oil and gas companies need to change their mindset to take responsibility for carbon dioxide. Green campaign groups need to accept that transformed fossil fuel extractors can be part of the solution.

Fringe benefits

Cleaner air: carbon capture from chimneys could also remove other local air pollutants.

More competitive renewables: increasing the price of fossil fuels would make zero-carbon energy sources cheaper by comparison.

New jobs: credible CCS will be a huge business needing many thousands of skilled workers. The jobs would be highly compatible with existing roles in the fossil fuel industry.

34
A Concrete Answer

'We produce enough concrete every year to build a six-lane highway to the moon.'

These are the words of Colin Hills, professor of environment and materials engineering, who then surprises me with a CO_2 gas canister, a plastic drinks bottle, some waste ash and a little water. He tips the ash in the bottle, adds the water, squirts in some gas, screws the lid on and shakes it up. Within seconds the bottle has imploded, crushed to a fraction of its former volume. And it is pretty hot.

'The CO_2 has been absorbed into the ash to create calcium carbonate – in effect limestone. It's a cost-effective way of reducing the cement industry's climate impact.'

And what an impact: making cement and bulking it out for concrete contributes around 7 per cent to our greenhouse gas emissions, rivalling steel as the single biggest industrial cause of climate change. Colin Hills is one of a growing band of scientists, engineers and entrepreneurs who are trying to shift it off the environmental naughty step because, despite the aspirations of wood reported in Chapter 31, our appetite for cement shows little sign of collapsing.

Cement and its combination with sand or stones to make concrete is one of humanity's great inventions. It's been used for millennia – it shaped the Roman Pantheon still standing two thousand years on. It is strong, versatile, durable and cheap. Just about every country has a cement industry. It has built our civilisation but cement's punishing carbon emissions now threaten it: a result of both the core chemical equation and the energy required to make it. The key ingredient in cement is limestone, calcium carbonate, which is baked in kilns at around 1,500°C (2,732°F) to break the carbon bond yielding calcium oxide and unwanted carbon dioxide. About one third of the CO_2 comes from the heating process, the remaining two thirds from the actual chemical separation. It all tends to go up the chimney.

A conceptually simple solution would be to use renewable energy for the heat and carbon capture and storage (CCS, discussed in Chapter 33) for the emissions. This may well become part of the solution but faces big hurdles: it is difficult and expensive to reach such massive temperatures without burning fossil fuels and a CCS industry of sufficient scale is decades away. What about treating the carbon dioxide as a useful raw material, not a waste? That was Colin Hills's insight in the 1980s: 'I started messing around with CO_2 at college. I thought it was incredibly good fun. But the professor told me off saying, "Go and do something useful." It's now very useful. Its time has come.'

Colin Hills's process is not changing the way cement itself is produced but using that industry's waste CO_2 to make artificial rock. That permanently locks up the troublesome gas. His spin-off company Carbon8 Systems (C8S) now makes and uses that rock as aggregate for building blocks. Carbonation – the reaction in the bottle described earlier – is the key.

Smart readers may well have noticed some chemical symmetry above: the equation at the heart of cement manufacture was reversed in the tabletop crushed bottle demo. Start with limestone and, after two reactions, you are back with limestone. Colin and his team discovered that a wide range of ashes and dusts left over from industrial processes would mineralise in the presence of CO_2, in effect, creating rock. It's an industrialised version of what molluscs do to make shells, which are then laid down over millennia to form limestone in the first place.

Most of the big cement companies claim to be working hard to cut carbon but, in the absence of a punitive carbon tax, any abatement technologies need to be as cost effective as possible. Carbon8's answer is to use some raw materials that are cheaper than free: it is actually paid to remove them. Air pollution control residue (APCR) is a fine dust of chemicals that is cleaned from the chimneys of waste incinerators or other dirty exhausts. It is highly alkaline, toxic and normally landfilled at great expense. Carbon8 gets a cheque for diverting this dust from a hole in the ground and then exposes it to the CO_2 flowing out of a cement kiln. The two ingredients combine, and the carbonation produces a strong aggregate for use in road stone or building blocks. This carbon capture and *utilisation* system, now containerised for ease of deployment, has been tested in Canada and commercially installed in France. So far it only absorbs a fraction of emissions but the principle is proven and commercially robust.

Other innovators are nibbling away at parts of concrete's size 12 carbon footprint. California-based Blue Planet Ltd uses similar mineralisation chemistry as Carbon8 but focuses on using the exhaust from power plants to create an aggregate to be used in concrete, locking up CO_2 and avoiding emissions from quarrying fresh rock. Its synthetic stone is in the runways of the recently extended San Francisco airport. Across the USA, New Jersey-based Solidia has developed ways to lower the required temperature of cement kilns and produce a cement that solidifies by reacting with CO_2 rather than water. The cement giant Heidelberg is trying to adapt the way it preheats the limestone to make it easier to capture pure CO_2 for use or geological storage.

As cement companies, or their customers, are prepared to pay more to shrink their carbon footprint, more solutions come into play. Colin Hills wants to combine his carbonation skills with another carbon-removal technology for an even bigger climate win. A technique called 'bioenergy carbon capture and storage' (BECCS) is one of the big ideas for future climate stabilisation. The idea is to grow trees to absorb carbon, burn them for useful energy and store the CO_2 underground. It could lead to a potentially global scheme involving massive forests and giant burners. Colin has his eye on all the resulting ash. It can be used to substitute up to 10 per cent of cement in concrete and also combine with CO_2 to form useful, stable minerals once again.

Just as digital companies are disruptive and footloose, cement companies have been solid and traditional. Their core processes are ancient and our buildings stand tall thanks to their time-proven techniques. But Colin Hills says their foundations are now shaking: 'These businesses like certainty. But they are going to be made to change or they are going to die. Yes, it will cost money but money is just an expression of will.'

Desirable destination

Reducing the climate impact of the cement industry by two thirds by 2050, cutting 5 per cent off our greenhouse gas emissions.

How to get there

Innovation: a vigorous scientific assault on the energy demand of concrete, improving CO_2 mineralisation and use technologies.

Taxation: building regulations requiring low carbon concrete usage and a carbon tax would be an instant accelerator of change.

Reduction and substitution: better design needing less cement and substation with lower carbon alternatives such as wood or quarried stone.

Fringe benefits

- Reduction in hazardous waste.
- Reduced quarrying and transport of aggregate.

35

Just Suck It Up

We have spent centuries putting too much carbon dioxide into the air from machines such as steam engines, cars, planes and power stations. Can we now invent a machine to pull it out again? Yes, and it's called 'direct air capture' or DAC.

On a plain between the snow-streaked flanks of Iceland's interior sits Climeworks's latest direct air capture plant. It looks like a bank of air conditioners the size of shipping containers and it's removing an amount of CO_2 equivalent to that emitted by 600 average Europeans. It is one of a handful of plants built today which, in 30 years' time, could grow into an enterprise as big as today's oil and gas industry.

'The core idea is to capture carbon dioxide from the air and permanently store it underground. The science and the principles have been around for 100 years. The art is to be cost efficient and energy efficient,' says Christoph Gebald, co-founder and CEO of Climeworks. The company is one of the major players in DAC, with plants in its home country of Switzerland before expanding to Italy and Iceland. In its modular system, fans blow air across a filter material that traps the CO_2. The key ingredients in the filter are strongly alkaline granules that attract the mildly acidic CO_2. Gradually the surface of the granules becomes saturated, closer to pH neutral and less effective. At that point, the granules are heated to 100°C (212°F) and they release pure CO_2, which is drawn off and captured. It must then be used or stored. Climeworks provides the technology for the actual CO_2 extraction from the air but is dependent on two vital industries from the outside: renewable energy and reliable carbon storage. In Switzerland, its initial plant ran on waste heat and electricity drawn from an energy from waste power station and it supplied the CO_2 to Coca-Cola for fizzy drinks. This is less than perfect in climate terms: the energy from the waste plant emits carbon and so do those gassy sodas. But they were both important 'bridge' partnerships to get the technology off the ground and prove it works. The future is the Icelandic model: zero-carbon energy, permanent carbon storage.

In Iceland, thc Climeworks plant is next to a geothermal power station. Iceland is famously volcanic because it sits on a fissure in the Earth's crust, allowing lava closer to the surface. That means you don't have to pump water down far before it turns to steam for generating electricity. Also, Climeworks's system uses 80 per cent of its energy as heat and just 20 per cent as electricity, so being close to hot rocks helps with that too. But what to do with the CO_2? In Iceland Climeworks has partnered with a company called Carbfix which dissolves the CO_2 in water and pumps it underground, where it mineralises in contact with the basalt bedrock (a similar process is mentioned in Chapters 20 and 34). It is turned into stone and will remain so for millions of years. Away from Iceland, Climeworks is looking to partner with CO_2-storage schemes planning to pump the liquid gas into former oil and gas fields in the North Sea. Climeworks's competence lies in the retrieval and concentration of CO_2, not the energy provision and final storage.

The cost of any proposals to reduce global warming are compared using dollars per tonne of CO_2 cut or captured. The price of the Climeworks system is high at $400 to $800 per tonne, against around $20 for tree planting and around $30 for solar or wind technologies. This will come down as the technology matures but is unlikely to rival the cheapest. Cristoph Gebald is unworried: 'What we are offering is unmatched permanence. We need to be planting trees but they can catch fire. We need to be fitting carbon capture direct to the chimneys of high emitters. But to stop damaging climate change we will *still* need to remove CO_2 from the atmosphere. In the 2020s, when discussing climate-friendly technologies, you should not be allowed to use the word "or". You should only be allowed to use the word "and".'

Climeworks is not the only player in the direct air capture game: there is also Carbon Engineering in Canada and Silicon Kingdom based in Ireland. Much of the money comes from other companies paying to offset their own emissions and many investors think the cost of DAC will

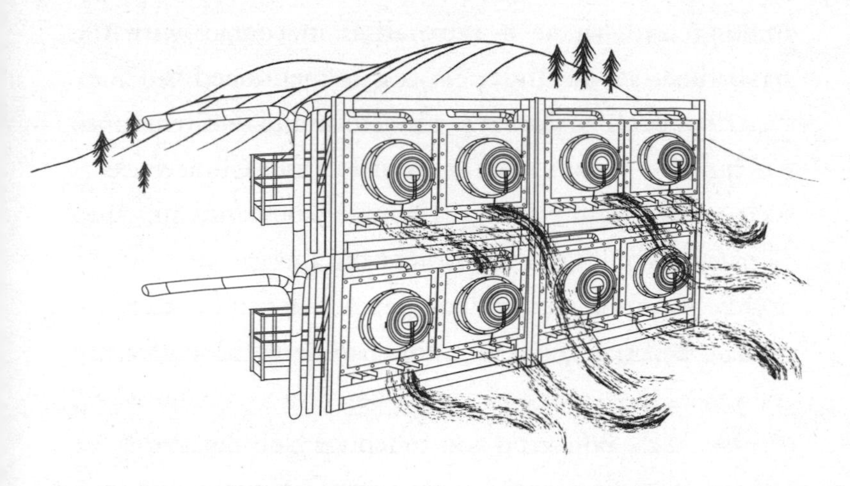

plummet as volumes increase, in the same way as we have seen with solar panels.

But, for the foreseeable future, there is no escaping the vast energy demand of DAC. Christoph Gebald has a plan for that, whereby every new solar farm or windfarm is built with on-site DAC: 'Then you can vastly increase the size and output of the renewables site, let's say a solar array with seven times what you need for the grid. On dull days you can still deliver enough electricity to the grid but on bright and sunny days you can power up the DAC devices. We can handle fluctuating energy inputs very well. CO_2 is everywhere, so we can be anywhere there is an energy source. So, my dream is to make DAC a catalyst for renewables – we aren't power storage, we are a power sponge.'

Despite its current high cost there is enormous interest in this technology. The scientists are clear that halting climate change doesn't just mean ending emissions but cleaning up what we've dumped in the atmosphere for generations. Political support is growing, especially from the US government under Joe Biden, and parts of the fossil fuel business see potential to use direct air capture to clean their waste. With this appetite Christoph forecasts that Climeworks will grow ten times bigger every two years. It may be an ambitious prediction, but this would mean, by the late 2020s, it could be absorbing 5–50 million tonnes per year – the upper end would be close to the emissions of New York. But could too much faith in our ability to remove carbon from the air undermine our efforts to stop polluting in the first place and give polluters a moral licence to continue operation? For me, it's a small risk as the cost is so big and, for at least a decade, the potential so small that no government serious about climate change is going to wait for DAC to get us off the hook. As Christoph said, it's about 'and' not 'or'.

Desirable destination

Capturing and storing the excess CO_2 in the atmosphere after we have decarbonised electricity, transport, heat, industry and agriculture as much as possible. Sharing the job with natural carbon stores such as trees, oceans and geological weathering.

How to get there

- Reducing the cost and energy demand, while increasing the efficiency of the direct air capture technology.
- Reliable, proven geological CO_2 storage.
- Public acceptance of higher energy costs or a climate tax to pay for the mass deployment of direct air capture.

Fringe benefits

New jobs.

36
Clean Steel

As plastic is the fabric of today's consumer world, so steel remains the stuff of manufacturing. With buildings, transport, machines and infrastructure, we live in a steel-framed society, and our civilisation would literally collapse without it. Unfortunately, steel has the biggest carbon footprint of any industrial sector, pumping out around 8 per cent of human-made CO_2 emissions. Along with the mastery of steam, smelting iron was the driving technology of the Industrial Revolution: for better or worse, it got us where we are today. So, can we create zero-carbon steel?

It is being done. But before coming on to where and how, it's crucial to understand the steelmaking process: mining the iron ore, crushing the ore, transporting the ore, smelting (essentially extracting pure iron from the ore) and processing into steel products. All these steps, especially the smelting, are very energy intensive.

Climate-friendly steel is being forged in the northern Swedish town of Luleå. A mining company, an energy company and a steelmaker have come together in a joint venture to form HYBRIT – Hydrogen Breakthrough Iron-making Technology. It is currently making 1 tonne of fossil-free steel per hour but has big ambitions. Martin Pei is the chief executive: 'If we make steel without carbon footprint, then steel becomes the enabler for the future sustainable society. It is extremely important for mankind to continue to develop. We need to solve this emission issue and we are quite confident that we are very close to a solution.'

Let's start underground. The HYBRIT project's mining partner LKAB has already switched to using electric power instead of fossil fuels in the lifting and crushing processes; diesel-powered trucks and loaders are the next target.

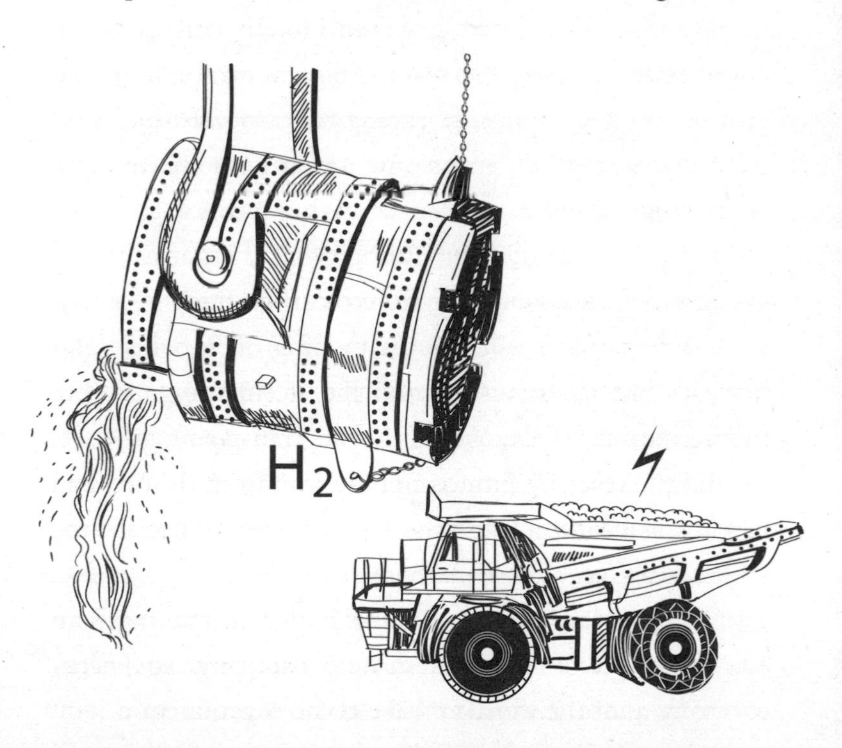

But the really tricky part of the process to decarbonise is the smelting, as it's not simply about using cleaner fuels as an energy source. You have to find something to replace carbon in the fundamental chemical equation of producing steel. Iron ore is mostly iron oxide, what we commonly see as rust, mixed up with other impurities. Efficient separation of pure iron from that ore using coking coal is so fundamental to our world that the place where it first happened, Coalbrookdale in the English West Midlands, is known as the birthplace of the Industrial Revolution. Mix the iron ore with carbon-rich coke at an extremely high temperature and the reaction that follows separates the oxygen from the iron, and the oxygen bonds to the carbon. This is what happens in a blast furnace and accounts for 85–90 per cent of the carbon dioxide emissions from ore-based steel production. In effect, iron oxide and carbon *in*, iron and carbon dioxide *out*.

The trick is to find something other than carbon that can tempt the oxygen away from the iron. Step up hydrogen. The hydrogen will react with the iron ore in similar furnaces but the resulting emission is water – H_2O, not carbon dioxide – CO_2.

But the resulting iron is not liquid, as it would be in a furnace, but spongy and requires further melting, mixing with scrap iron and refinement before it can make steel products. All this requires more energy. Steelmaking is an epic energy user. In total it demands 8,000 TWh, equivalent to one third of the world's total electricity production.

Often a steelworks uses more energy than the neighbouring town. This is why the steel plants grew on top of coalfields in Britain, France, Germany, the USA, China, pretty much the world over, and why a key partner in Sweden's HYBRIT project is Vattenfall, a company focused on renewable energy. Carbon-free electricity is needed not just to power the plant but to produce the clean hydrogen. It is estimated that producing all the world's steel with hydrogen would demand all the energy from three times the number of wind turbines the world has today. To then create all that hydrogen demands increasing electrolyser capacity from 1 gigawatt today to 650 gigawatts. All this is technically doable but little short of a new industrial revolution and will change the geography of steel production. Future, low-carbon steel plants may well arise near low-carbon energy sources such as windfarms, hydroelectric dams, huge solar fields or nuclear power stations. Changing where such a big industry happens creates winners and losers, making such change politically uncomfortable.

So, will this change actually happen? Martin Pei of HYBRIT says the first step is proving its system is robust and can make quality products around the clock. He welcomes other big manufacturers in Germany and India also investing in the technology as it shows confidence that the whole industry will move this way. But he does acknowledge that for quite a few years, zero-carbon steel will be more expensive to make than the current dirty version. Ensuring its success demands a high carbon price to penalise traditional producers, customers paying a premium for zero-carbon steel and even temporary tariff barriers to protect clean steelmakers from 'dirty' imports.

As tall orders go, carbon-free steel by 2050 through using hydrogen and green electricity is a pretty lofty one but there are some other ideas for shrinking our metal carbon footprint. One is to ensure blast furnaces are equipped with carbon capture and storage capacity, a big new industry in itself – as discussed in Chapter 33 – but maybe easier in the short term than revolutionising the actual steelmaking process. The other is to use less steel.

Every year about 1.8 billion tonnes of steel are produced. That is about one fifth of a tonne per person on earth, about three times average bodyweight. Do we really need to be using all that steel in the first place? The main markets for it are buildings, transport and industry itself, all roughly taking one third each. But by improving design to require less weight, developing longer life products, reusing components, recycling more and substituting other less polluting materials, it would be possible to cut steel production. At the moment, steel is too cheap to drive those efficiencies, but zero-carbon steel would be more expensive and, as with energy or food, paying the full environmental cost of production means we would use it more smartly.

Desirable destination

Eliminating the 8 per cent of greenhouse gas emissions that come from steelmaking by designing to use less steel and making the remainder with hydrogen and electricity or blast furnaces fitted with carbon capture and storage.

How to get there

Technology: scaling up the clean steel process to reduce the cost.

Hydrogen production: massively increasing the volume of green hydrogen produced from renewable energy or nuclear power.

Policy: regulation to insist locally produced steel or imports meet steadily stricter carbon credentials.

Fringe benefits

- Less local air pollution near steel plants' coking-coal mines.
- Development of the hydrogen economy.

Waste

37

Climate Time Bombs

The gases used to keep us and our food cool are a climate time bomb. If they escape into the air, they'll warm the atmosphere by around 0.5°C and, given our total global warming to date from pre-industrial levels is about 1°C, that is a huge figure. Some experts have calculated that stopping this could be the single most significant solution to climate change. So how do we keep our bodies, food *and* the planet chilled? Start by tracking down, containing and destroying the dangerous chemicals. María José Gutiérrez leads one such group of hunters. In her high-school yearbook, written a few years back, it said, 'Maria will save the world.'

Fridges work by circulating a refrigerant chemical through pipes and a compressor, letting it change from liquid to gas to create the chilling effect. In the last century, the most commonly used gases were CFC (chlorofluorocarbon) and HCFC (hydrochlorofluorocarbon) but in the 1980s it was discovered their release was causing the hole in the ozone layer over the North and South Poles. The atmospheric shield against harmful ultraviolet radiation from the sun was shrinking. In a remarkably rapid and coordinated international response, the Montreal Protocol, signed in 1987, required CFC and HCFC production and use be phased out and the ozone hole is now closing. This was good news for climate change too as both the banned chemicals are potent greenhouse gases; less welcome is that the initial replacement HFC (hydrofluorocarbon) has a global warming potential 1,000 to 9,000 times that of CO_2. The insulating property that makes these gases so good for fridges is exactly what makes them such a perilously effective blanket for the world. When you combine refrigerant gases in use and what is in store, it adds up to a frightening stockpile.

María José Gutiérrez, from Costa Rica, works for Tradewater, an American company committed to the discovery and destruction of refrigerant gases before they leak out. She likens her team to 'Ghostbusters, but for fridge gases' and, working across South and Central America, they're known as the RefriCazadores – 'Chill Hunters'. 'I have seen the effect of climate change in my region and I want to make a real, measurable impact. I want these gases out of the atmosphere.'

The team are after full gas tanks, intact fridges or industrial chillers often stored in old warehouses and waste-disposal sites. Once found, and a deal is struck with the owner and local authorities, they can be removed and the gas destroyed in an incinerator. But it is rarely that simple. These chemicals exist in a legal grey area, so stocks are often hidden, owners may hope to sell them in the future or sometimes simply the scrap iron value of the canister means the gas is vented and the metal sold on. In the worst cases, the gas from a single 15-kg cylinder can have the same global warming potential as driving 54 cars for a year. Tradewater cleaned up a site in Ghana holding 771 cylinders, removing at a stroke the climate change impact of 27,600 cars. Sadly, sometimes the team arrive too late, finding only punctured tanks, corroded pipes and gases now doing their worst. These gas canisters are weapons of mass destruction for our climate, just waiting to go off.

A successful seek-and-destroy mission demands detective work, diplomacy and sometimes deep pockets. Tradewater's money comes from offsets where companies or individuals will compensate for their carbon emissions, from say manufacturing or flying, by paying for climate-friendly projects elsewhere.

At the time of writing, Tradewater's gas recoveries from around the world have prevented the equivalent of 5 million tonnes of CO_2 from reaching the atmosphere but María José Gutiérrez isn't satisfied: 'We are only scratching the surface. There is so much more out there. There are some people that have this gas and want it disposed but others that don't want it disposed so we need to be creative and make sure we get it all. It's challenging work but it's very exciting as well because of the tangible effect it can have on climate change.'

Staunching the flow of refrigerants from the past is only part of the solution: we need to control and replace the

gases we use currently. Once again, international cooperation provides some refreshing optimism. An amendment to the Montreal Protocol, thrashed out in the Rwandan capital of Kigali, committed richer countries to begin phasing out HFCs back in 2019 with lower income countries following throughout the 2020s. Unlike recent climate accords, it is mandatory not voluntary. But the job is huge and growing as demand for air conditioning and chilled food storage grows in the developing world. More climate-friendly alternatives to HFC – propane, ammonia and, ironically, carbon dioxide – are being introduced but they require new equipment, training and, in the case of propane, new safety protocols as it's an inflammable gas.

The United Nations Industrial Development Organization (UNIDO) is coordinating these efforts, especially in lower income countries. It has backed climate-friendly chillers in a supermarket in Jordan and HFC recovery training in Ecuador, and promoted alternative refrigerants in the air conditioners of the Gambia. Overall, UNIDO has 20 such programmes in 18 countries and in 2019 alone saved the equivalent of 25 million tonnes of carbon dioxide. Not the biggest, but a personal favourite, is Brazil's first climate-friendly beer cooler. UNIDO worked with a locally owned business from Ribeirão Preto to produce a unit able to cool 20 litres of beer an hour while reducing the climate impact of the refrigerants by a factor of 1500. I'll drink to that.

Desirable destination

Reducing average global temperature rise by 0.5°C by 2100 by avoiding the emission of refrigerant gases.

How to get there

International politics: implement and enforce the Kigali Amendment to the Montreal Protocol, which is a legally binding agreement to phase out hydrofluorocarbons. As of spring 2021, 118 states and the European Union were signed up; the USA and China had promised to ratify.

Financial: support the work of Tradewater and other organisations through climate offset funding to secure stockpiles of old gas tanks.

Fringe benefits

Recycling of aluminium and steel in fridges, air conditioners and gas tanks.

38
Liquid Air

Food waste is one of the biggest and dumbest causes of climate change.

We need food. We love food. Yet we chuck food away. Food is essential for the body, delightful for the mind and plentiful in landfill.

One third of what is grown is never consumed. Its growth produces greenhouse gases and its rotting releases more. Around 8 per cent of human-induced climate change comes from food we waste: about the same as emissions from India and Germany combined. Without it we could let one third of our farmland – an area the size of Russia, the world's biggest country – return to nature. That could have enormous carbon-removal potential and, combined with avoided emissions from less rotting food, would be a big bite out of climate change.

Food goes uneaten throughout the world but there is a stark difference in the way it's lost. In richer countries, most is wasted after it reaches our homes and addressing that is mainly a labelling, cultural and behavioural challenge, which we'll come to later. In poorer countries, most of the losses happen before food even crosses the threshold, as it rots on the way. But not for much longer if a British garage inventor has his way.

Peter Dearman is a genius destined to become a household name like Ford and Hoover. He really does still work in a garage. It's on a terraced street, behind a car missing a wheel, in the English county town of Bishop's Stortford. Lift the overhead door to reveal a jumble of lathes, drills, milling machines and gas bottles, lit by a couple of naked lightbulbs hanging on their wires. This is where he invented and improved an engine powered by air: the Dearman Engine. Air is 80 per cent nitrogen, and once compressed and cooled it can be stored and transported in tanks like fuel. When you open the tank valve, the nitrogen expands 700-fold to become gas with considerable force – enough to push a piston and drive an engine. This engine can generate electricity to run a refrigeration unit, and emits no carbon but lots of *cold* nitrogen. That combination of pollution-free electricity and chilling exhaust is a perfect for climate-friendly cooling.

In low-income countries, the main reason that so much food perishes before it gets home is the absence of a 'cold chain' – the network of refrigerated trucks and warehouses that keeps food fresh and is taken for granted in the richer world. Most of this is powered by diesel-fuelled chillers emitting both greenhouse gases and unhealthy, street-level air pollution. Expanding that across the world would be a climate change disaster but Dearman has the solution.

His engine has been taken out of the shed and refined by Clean Cold Power, a London-based company with big ambitions to stop cooling our food from heating the world. Its commercial director Khaled Simmons tells me there are 2.5 million diesel refrigerated trucks worldwide and this could quadruple in the next decade. The fuel consumption of the refrigeration unit is, on average, about 20 per cent of what it takes to drive the lorry down the road and the fridge motor keeps going when the vehicle is parked. But chugging away in Khaled's yard is a full-size refrigerated lorry trailer cooled only by nitrogen. The first commercial market is likely to be California where tough air-quality regulations will steadily drive out fossil-fuelled chillers. He hopes, thanks to his engineers working tirelessly to cut cost and improve efficiency, that nitrogen cooling will edge out diesel in much of the existing fleet: 'Liquid nitrogen engines have this perfect win–win for zero-carbon cooling: an engine that runs on air with an "exhaust" which is sub-zero.'

But the really big bite out of climate change would come if this technology were widely adopted in poorer countries to reduce food waste. It is challenging, though, as diesel, its dirty rival, is widely available. But then, so is air … and it's free. Turning air into liquid nitrogen is not super hi-tech but it is power hungry so needs to be done with renewable electricity, like solar or wind for lifecycle carbon savings.

In wealthier countries, such as the UK, it is consumers who are wasters. Roughly one third of food we buy is binned. Between the farm and the shop, we have a very efficient, albeit carbon-intensive, cold chain that stops food spoiling en route. But when it gets into homes, that's when the rot sets in. In Europe, just over half the food wasted is discarded at home; in the UK the figure is 70 per cent; in the USA it is lower at 43 per cent partly because Americans eat out more. The simple truth is that we can afford to throw food away: we don't *need* to eat a slightly soft banana or curling slice of bread when there is something newer or nicer on the shelf. Also, transforming tired food into a tasty treat takes confidence, knowledge and a little time: all ingredients said to be in short supply in our kitchens. But does this feel like a valid excuse for binning it when you consider food waste results in 20 million tonnes of CO_2 emissions every year in the UK – equivalent to around 10 million cars – when most of it is still edible? The campaign group WRAP (Waste and Resource Action Plan) went through the bins and found nearly 70 per cent of the food within had not gone off.

An effective yet completely unpalatable solution would be make food expensive and thus more valued. It won't happen, at least not deliberately, as nearly all politicians trumpet the virtue of 'affordable' food and, without better support, some of the poorest families might struggle to eat enough. So how else can we reduce our waste? Manufacturers and supermarkets have developed smaller products for smaller family units; volume promotions such as 'Buy one, get one free' are becoming rare; 'Use by' dates not merited by food safety are slowly vanishing. Campaigns from charities, government and celebrity chefs have all demonised waste and promoted leftovers. And there has been some success. Since 2007 food waste has reduced by over one quarter in the UK and figures approaching that have been achieved in much of Europe. Andrew Parry, from WRAP, hopes waste can be driven down further with a change in tactics. The message had been focused on how food waste hit us in the pocket as it was felt most consumers did not yet understand the environmental impact of food production. But now, with younger people caring more about climate change, that is going to be in the 'shop window' of the waste-cutting message.

Morality is never far from the food-waste debate as nourishment is the stuff of life. Many religions celebrate harvest and believers thank a deity before eating. In most conflicts, notably the Second World War, wasting food was considered a deeply shameful act. The terrible and highly publicised Indian and African famines of the twentieth century were a further reminder not to squander the food on our plates. It may take the combined moral force of preventing climate change and cutting food waste to drive the behavioural and technical changes needed to ensure we eat what we grow.

Desirable destination

- Direct carbon savings from decarbonising refrigeration is around 0.1 per cent.
- Reduce food waste and plant the land made available with trees, saving around 5 per cent of emissions by 2050.

How to get there

- Continued improvement and deployment of liquid air technology and infrastructure.
- Tighter air pollution regulations.

Fringe benefits

- Improved urban air quality from the removal of polluting transport refrigeration units.
- More land made available for nature.

39

Grow More, Pollute Less

Close to half of the world's population exists today thanks to the discovery made by German chemists Fritz Haber and Carl Bosch at the start of the twentieth century. They combined nitrogen and hydrogen to make ammonia – the key ingredient of artificial fertiliser, which has enabled food production to boom and support a global population now approaching 8 billion. That's over four times what it was when they invented the Haber–Bosch process. But making and using all that synthetic fertiliser has come at a price: polluted water, degraded soils, less nutritious food and, now top of the list, massive global warming. Thankfully, the meeting of two fathers at the school gates in Yorkshire might allow us to keep the bumper crops but lose the punishing side effects.

'I've had good results capturing carbon dioxide with treated natural fibres. I know you do a lot with money. Can we get together to takes things forward?' Thus spake industrial chemist Peter Hammond to financier Pawel Kisielewski as they dropped off their children. They are now making ultra-low carbon fertiliser from waste products and believe there is no technical reason why all fertiliser couldn't be made this way. One of their first partners is the food and drinks company PepsiCo and one of the first products to reap the climate-friendly benefits is potato crisps. By the end of 2022 they might be sporting a net-zero sticker.

Traditional chemical fertiliser drives climate change in both its manufacture and use. The raw ingredients are nitrogen, which comes from the air, and hydrogen, which is obtained by separating it from the methane molecule – CH_4. This not only requires plenty of heat but the reaction itself leaves you with CO_2. Then you combine the hydrogen and nitrogen under great temperature and pressure, which requires yet more energy. Once the resulting ammonia is spread on the field, up to 50 per cent never helps the plant and simply evaporates as nitrous oxide: a greenhouse gas with a potency nearly 300 times that of carbon dioxide. All that adds up to the carbon footprint from fertiliser, which is about 2.5 per cent of our total global warming, up there with aviation. Or, seen another way, around 4 tonnes of CO_2 equivalent for every tonne of fertiliser.

Peter Hammond and Pawel Kisielewski formed the company CCm Technologies to make fertiliser from waste materials – fibrous organic matter, ammonia from food waste or sewage, and carbon dioxide from industrial chimneys. The resulting fertiliser is high in ammonium nitrate and organic matter; it can be tailored to suit the crop or country, from rice in Thailand to potatoes in Britain; it's no more expensive than the conventional stuff; and as a pellet or granule it can be spread with the same kit.

Peter says there's a bit of Doc Brown's engine in *Back to the Future* in their approach: 'Conceptually, it's similar to what we used to do on an organic farm – using the waste to help the crops grow – but optimised for the twenty-first century. Food leftovers, crop residue, compost, animal dung, human sewage and, crucially, carbon dioxide all combined. It's recycling those nutrients, it is genuinely circular and feels very good.'

Their fertiliser-manufacturing plants can find a home anywhere along the food chain – from the growing stage through to after the food is eaten. Let's start at PepsiCo's Walkers Crisps factory in Leicester. Here CCm Technologies has an anaerobic digester fed on waste potato peelings, which provides three quarters of the factory's electricity but it also yields CO_2 and what's left over after digestion, known as digestate 'cake'. These are combined by CCm along with further CO_2 from chimneys and ammonia from nearby wastewater treatment works to create fertiliser. This in turn is then used on the fields to grow potatoes. It is delivering a massive bite out of the carbon footprint of this bestselling salty snack. Retailers, other food companies and countries across the world are taking notice, says Pawel Kisielewski. 'Our technology can be zero carbon or even carbon negative. It's not going to cost you money and will help clean up your waste. It is a big claim … but, yes, I think we can replace a huge proportion of the world's conventional fertiliser.'

Another favoured habitat of CCm Technologies is the water treatment works. At a sewage works near Birmingham it is turning what's flushed out of the city into fertiliser: from sewage farm to arable farm. Wastewater has high concentrations of ammonia and phosphorous, which is a massive pollution headache for the water companies but can be extracted to provide the key ingredients for fertiliser. The water company has an on-site anaerobic digester fed on sewage solids, which once again yield digestate and CO_2. For CCm Technologies, ammonia, organic material and CO_2 are the holy trinity. Pawel Kisielewski: 'I'm in my early 60s and I've never been so emotionally engaged in a business. When it comes to climate change, agriculture has been behind the curve and we can help change that. This is for real.'

This is one of those climate change solutions with enormous fringe benefits. Many farms have let their soil's organic matter decline and this fertiliser helps put it back. In the long run this means less fertiliser required, more nutritious food and even better soil water retention, which can be a huge deal in areas prone to drought or floods – water is less likely to run off and inundate the nearby town. Also, in many countries with highly intensive livestock systems, such as parts of the USA or Holland, so much animal dung is creating a water-pollution crisis. Lakes and rivers contaminated with effluent are prone to blooms of algae, which suck up the oxygen and kill the wildlife. All that manure is the perfect raw material for fertiliser.

Greening the fertiliser industry while keeping food on the table is an epic task. But I think CCm Technologies, and others like it, can succeed because they don't need vast new dedicated factories. The idea and the technology can spread, almost unseen, throughout our existing food and waste infrastructure, like a super-serum turning a climate culprit into a climate champion.

Peter Hammond is quietly impatient: 'It is the easiest way to knock down greenhouse gases and sort out some problem wastes. It is here, you can roll it out quickly. Why not get on with it?'

Desirable destination

Eliminate the 2.5 per cent of our greenhouse emissions that comes from the production and use of synthetic fertiliser.

How to get there

- Deployment of fertiliser production technology wherever waste is treated: food waste, animal waste and human waste.
- Climate labelling of food to encourage the buying and production of low-carbon crops and livestock.
- Reforming regulations so the use of waste is encouraged instead of penalised.

Fringe benefits

- Better food nutrition.
- Better soil health.
- Better farm wildlife.
- Less water and air pollution.

Acknowledgements

This book began life as a radio series and the team at BBC Radio takes much of the credit. Firstly, Mohit Bakaya, Controller of BBC Radio 4. He backed the idea of a climate change solutions series when I first pitched it to him and gave it eight weeks of prime radio time along with all the podcast support. My friend, Alasdair Cross, produced the series and advised me on the book. As long-term collaborators we make a good creative team in crafting the sound, tone and content of the radio show. Much of this found its way into the book. Other friends from the radio posse were also invaluable: Anne-Marie Bullock with her relentless energy and dogged pursuit of news outlets, Sarah Goodman for inspired and thorough research. Radio 4 commissioner, Daniel Clarke and BBC Executive Editor, Dimitri Houtart provided support and editorial guidance throughout.

Although not quoted in the book, Dr Tamsin Edwards's expertise proved essential. A Reader in Climate Science at Kings College London and fellow of the Royal Geographical Society (RGS), she is my academic co-presenter on the radio show. Her warmth and knowledge not only shine through on air but also helped to form my thoughts for the book. Thanks also to the expertise of RGS advisers and the Director Joe Smith.

On the literary side, thanks to my agent Patrick Walsh who patiently guided this novice through the publishing jungle and who (despite the best efforts of COVID) I managed to see at least once over a decent meal rather than a Zoom screen. At Penguin, I had great help from Albert DePetrillo, Bethany Wright, Laura Nicol and Daniel Sorensen. From initial discussions, through to formatting, editing, illustrations and publicity, they led this ingenue through writing his first book.

My family all played their part too. Number one son, Dougal, shaped my thoughts on how to present the scale of each idea. Number two son, Edward, produced some great social media films to accompany the radio series and number three, Samuel, advised me on the ideal length of each chapter. My wife, Tammany, is my invaluable sounding board for ideas: enduring my frequent babbling about the latest chapter and advising where I might have lost the general reader. My interest in the environment was sparked by my father John's tireless championing and protection of the Antarctic while my mother Peg, amongst so much more, taught me the value of food.

But my greatest debt is to the women and men featured in this book as their endeavours fill these pages and, probably more significantly, can save the planet.